Your own Eco-Electrical Home Power System

D. Fichte

Your own
Eco-Electrical Home Power System

D. Fichte

Elektor International Media BV
P.O. Box 11
6114 ZG Susteren
The Netherlands

British Library Cataloguing in Publication Data
A catalogue record for this book is available from the British Library

ISBN 978-0-905705-82-8

Prepress production: Kontinu, Sittard
Design cover: Helfrich Ontwerpbureau, Deventer
First published in the United Kingdom 2009
Printed in the Netherlands by Wilco, Amersfoort

099003-UK

Table of Contents

The Typical Home Electric Power System

Electric energy had become such an integral part of life in the 20th century that it is hard to imagine doing without it in the 21st century. Yet the high rate of energy use in the developed world is straining the supply rate at a time of 'peak oil', the stagnating production of oil in a world with an increasing energy demand. Not only are gas prices increasing, but so are most electric utility rates. While these trends develop, the sun continues to supply one kilowatt per square meter (that is, 1000 watts over each 10.76 square feet) when directly overhead on a sunny day. Wind in some places continues to blow as it has for lifetimes. Streams continue to run. Why not benefit from this available energy?

This book is for those who want to know more about how to provide electric power for a household off the utility distribution network or 'power grid'. In some places, such as the jungles of Central America, no utility electricity is available. Examples given are from a Central American house on a jungle hilltop. Yet the power-system designs presented here could just as well apply to a suburban residence in Melbourne, Houston, Toulouse, Liverpool, or Manitoba. The approach taken here is to explain typical systems, their components, and how to optimally select, or 'size' them. In addition, the usual power system problems (and some more subtle ones) are described and analyzed, with practical recommendations on how to solve them.

Energy systems, like any other systems, can be decomposed into the functions they perform. These functions are:

- Generation: power originates from a source, such as sun, wind, a stream or a fuel-burning engine.
- Storage: electricity is stored in some form - usually batteries - to supply electricity when the generating source(s) do not.
- Conversion: electricity comes in different 'forms'. The form provided by solar panels is not the same as the form that occurs at the electric outlet in the typical home.

The first step in understanding power systems are the basic concepts of electricity.

A typical home electric power system is shown as a block diagram - that is, as a system - below. A *system* is an interconnection of *components*, shown as blocks. The system is the whole; the components are the parts of the system.

The Typical Home Electric Power System

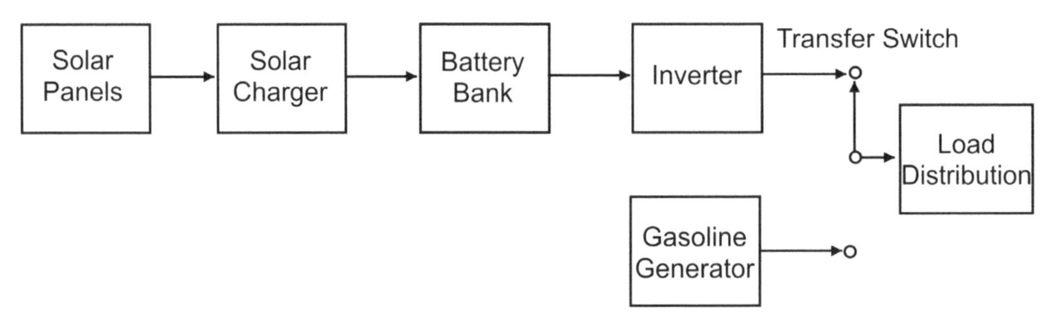

Solar panels are shown as the power source. To charge the batteries from the panels, a piece of electronics is needed, commonly called a *solar charger*. It converts from the panel voltage to the battery voltage as it controls the charging of the batteries. A good solar charger will not only charge the batteries for maximum battery life but will also use the maximum available power from the panels.

The batteries connect to an *inverter*, a second piece of electronics. It converts from battery voltage to the home distribution voltage of 120 V. For European use or higher-power equipment that operates from 240 V, both 240 V and 120 V outlets might be required from the inverter. The inverter output is then connected to a *transfer switch* which feeds into a circuit-breaker box (or *load center*), not shown. This switch selects the source as either the inverter or a backup supply, usually a gasoline or diesel generator. The transfer switch then connects to the house wiring and the various loads connect to it.

This system diagram will guide us in understanding how its various components relate to each other to achieve the desired functions of a home electric system. An additional block between the generator and batteries can be added: a *line-operated* or 'off-line' 120 V-powered battery charger that charges the batteries when the sun does not. Other additions are possible as are other electric power schemes, to be presented later.

Basic Electricity

This chapter explains the basic concepts of electric circuits and components. If you are already knowledgeable of electric circuits, it can be skipped. If you are not, I recommend a light read through it to become familiar with the concepts. Then refer back to it as a reference for the rest of the book.

Voltage, Current, and Resistance

Electric circuit quantities are analogous to fluid quantities. Voltage is like pressure and occurs *across* two places. Just as water pressure can be measured against atmospheric (gage) pressure or vacuum (absolute) pressure or between two elevations, voltage can also be measured between different points or *nodes* in an electric circuit. In water circuits, water can flow from a pressurized tank through a pipe and into the ground. This is an open circuit since the water does not return to the source (except globally). In electric circuits, the 'stuff' of electricity, which is called electric *charge*, flows in wires like water in pipes. The charge flow rate is the *current.* Sometimes the current is thought of as what is actually flowing, though electric charge is the electricity.

In water systems, less flow will occur through a small diameter pipe than a large one for the same pressure. The small pipe has a higher resistance to flow than a large pipe. In electric circuits, something similar happens, and the relationship is expressed as the most basic of all electric equations, *Ohm's Law*:

$$v = i \cdot R$$

where v is voltage, in units of volts, abbreviated V; i is current in ampères (or 'amps'), A; and R is resistance, in units of ohms, expressed by the symbol, Ω. A thousand ohms is written as 1000 Ω or 1 kΩ, where the unit prefix, k, stands for 'kilo' or 1000.

If a fan draws 1/3 A from a 120 V outlet, what is its resistance?

$$R = \frac{v}{i} = \frac{120\ \text{V}}{0.33\ \text{A}} = 360\ \Omega$$

The fan, as with many other loads, will not present a constant resistance to the power line. When starting, its resistance is much lower and the starting current is much higher. This must be taken into account when determining the 'size' of a system.

Two Circuit Laws

Two circuit laws, called *Kirchhoff's Laws*, describe essentially what goes on in electric circuits. Kirchhoff's Current Law (KCL) states that charge does not accumulate in nodes. If wires are connected together to form a node, as shown below, then the sum of the currents into the node must equal the sum going out, so that the net current is zero. That is,

$$i_1 + i_2 + i_3 = i_4$$

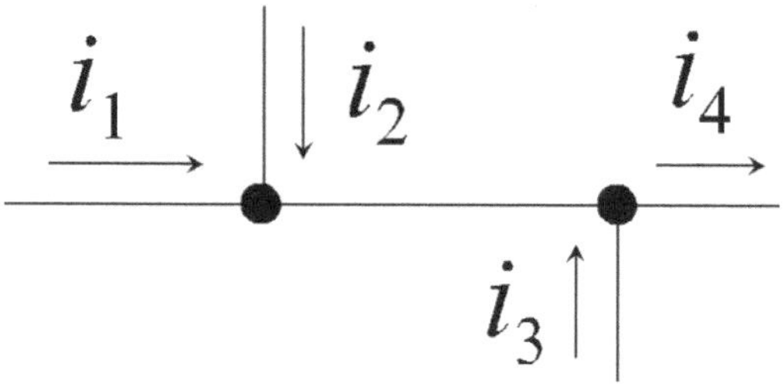

The basic point to remember is that current, which is moving charge, cannot accumulate in wires or in a node formed by them. The dots in the above fragment of a circuit diagram indicate connections of lines, which represent wires or other conductors.

Suppose that a battery charger powered by a generator is connected to a battery along with a solar charger. The solar charger puts out 35 A and the charger outputs 30 A. Then the total current into the battery must be 35 A + 30 A = 55 A. All current from the sources must go into the load, which in this case is the battery. None accumulates in the wires.

The second basic circuit law is Kirchhoff's Voltage Law (KVL). It states that the sum of the voltage drops around a closed-circuit loop must equal the voltage of the source. An example is shown below, where the circle is a symbol of a voltage source and the other symbols are of resistances, labeled R_1 and R_2. The closed loop allows current, I, to flow out of the positive (+) terminal of the 12 V voltage source, through R_1, then R_2, and back to the negative (−) terminal of the voltage source. If the loop were opened anywhere, the current would be zero. A broken or disconnected wire would open the loop and this is called an *open circuit.* The resistance of an open circuit is infinite.

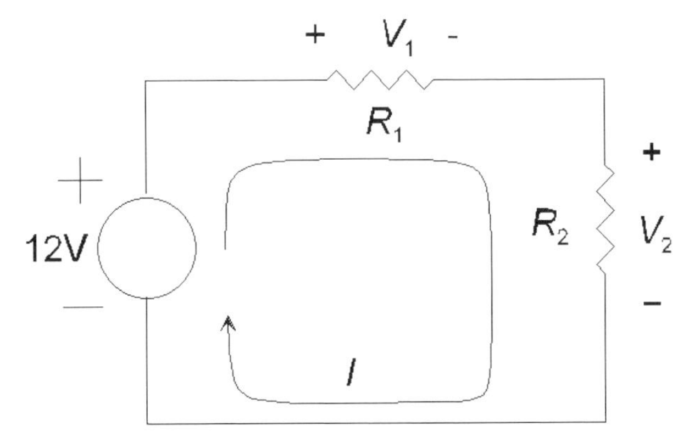

A *short circuit* is a wire or other electrically conductive connection where there should not be one. The resistance of an 'ideal' short circuit is zero, sometimes informally called a 'dead short'. For instance, if a wire were connected across the terminals of R_2, it would be short-circuited or shorted.

The current causes a voltage drop across the resistances, as shown. When KVL is applied to the above circuit, then

$$V_1 + V_2 = 12\text{ V}$$

We can find the values of V_1 and V_2 using Ohm's Law. The voltages are

$$V_1 = I \cdot R_1 \text{ ; } V_2 = I \cdot R_2$$

Resistances in series add, as in the above circuit. Therefore, the total resistance across the 12 V source is $R_1 + R_2$. Then using Ohm's Law,

$$I = \frac{V}{R} = \frac{12\text{ V}}{R_1 + R_2}$$

The 12 V source is applied across both resistors in series. Then the voltages across the individual resistances are

$$V_1 = \frac{R_1}{R_1 + R_2} \cdot 12\text{ V} \text{ ; } V_2 = \frac{R_2}{R_1 + R_2} \cdot 12\text{ V}$$

The series resistors form a *voltage divider*. The 12 V source voltage is divided among them proportional to their resistance. The above voltage-divider formulas can be used to calculate voltages of interest.

For example, suppose the 12 V source is a solar panel, R_1 is the resistance of the wire from the panel to a lamp and R_2 is the lamp resistance. Suppose $R_1 = 1\ \Omega$ and $R_2 = 9\ \Omega$. Then substituting these numerical values into the voltage divider formulas, we calculate the voltage, V_2, available across the lamp as 10.8 V and the voltage drop in the wire as 1.2 V. The wire and lamp current is 12 V/10 Ω = 1.2 A.

This example can be used as a template for calculating how much voltage will appear across a load when wire of a certain resistance is used to deliver power from a source. Wire size is ‘gauged’ in conductor (usually copper) diameter or cross-sectional area and given an American Wire Gage (AWG) number or, more commonly nowadays using metric measure, given in square millimeters (mm^2). The table on the following page lists resistance per length of wire for various sizes at room temperature (20 °C or 68 °F), the conductor diameter, and the maximum current for a current density of 450 A/cm^2, a value that is somewhat conservative. Wire suspended in the air or run straight in a pair can conduct a somewhat higher current density because it is better cooled. A change of 6 AWG numbers doubles or halves resistance.

As an example of table use, a pair of wires is run from a generator (the source) to a water pump (the load). The pump requires 2.08 A to run from a 120 V source. The pump voltage should not be lower than 110 V when operated. What size of wire is required to run to the pump over a distance of 1200 feet?

First, there are 0.3048 m/ft. Then

$$(1200\ \text{ft})\cdot(0.3048\ \text{m/ft}) = 365.8\ \text{m} = 0.3658\ \text{km}$$

The pump resistance, when operating at minimum voltage, is

$$R_L = 110\ \text{V}/2.08\ \text{A} = 52.9\ \Omega$$

Using the voltage-divider formula,

$$110\ \text{V} = \frac{R_L}{R_w + R_L}\cdot(120\ \text{V})$$

where R_w is the wire resistance. Solving this equation for R_w using algebra, the formula for maximum wire resistance is

$$R_w = \frac{V_g - V_L}{V_L}\cdot R_L$$

AWG#	Diameter, mm	Resistance, Ω/km	Maximum current , A (450 A/cm^2)
0000	11.68	0.1608	482
000	10.40	0.2027	382
00	9.27	0.2557	304
0	8.25	0.3224	241
1	7.35	0.4065	191
2	6.54	0.5128	151
3	5.83	0.6463	120
4	5.19	0.8153	95
5	4.62	1.028	75
6	4.11	1.30	60
7	3.66	1.63	47
8	3.26	2.06	38
9	2.91	2.60	30
10	2.67	3.29	25
11	2.38	4.137	20
12	2.13	5.209	15
13	1.90	6.964	12
14	1.71	8.280	10
15	1.53	10.43	8
16	1.37	13.18	6
17	1.22	16.58	5
18	1.09	20.95	4
19	0.948	26.39	3
20	0.874	33.23	2.75
21	0.785	41.89	2.5
22	0.701	53.14	2
23	0.632	66.60	1.5
24	0.566	84.21	1

where V_g = generator (source) voltage = 120 V. Substituting the numbers into the formula,

$$R_w = \frac{V_g - V_L}{V_L} \cdot R_L = \frac{120\text{ V} - 110\text{ V}}{110\text{ V}} \cdot 52.9\ \Omega = 4.81\,\Omega$$

Then the resistance per length is: 4.81 Ω/0.3658 km = 13.15 Ω/km. Because there are two wires, one in each direction of current flow, the effective length is doubled, and the wire must have a maximum resistance per length of half that, or 6.57 Ω/km. From the table, 12-gage wire is the smallest that should be used.

Reactance

Resistances dissipate power as a heating rate or convert it to some other form of power. Motors, for instance, convert some power to heat and most of it to mechanical power. All of the power drawn from the source is used.

Two other basic kinds of circuit *elements* are *capacitance* and *inductance*. Unlike resistance, these do not dissipate or convert power but store it instead and give it back later. These kinds of circuit elements are called *reactances*. Because power is returned, this can cause trouble for the source, especially if it is electronic such as an inverter. Reactance also has values in units of ohms (Ω), but they are not 'real' ohms as resistance ohms are. They are envisioned as being along another dimension, as shown in the plot below.

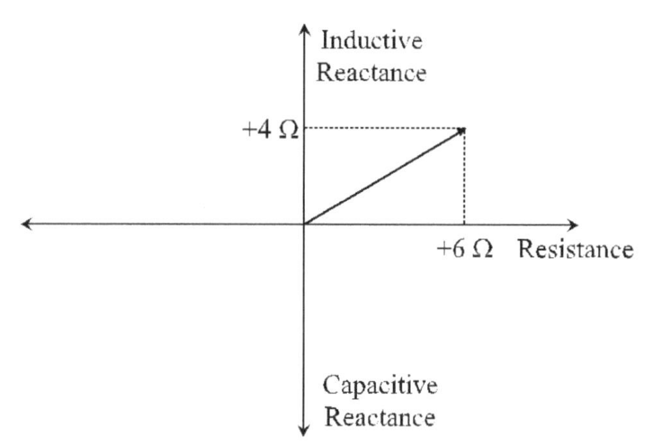

In this two-dimensional plot, resistance is plotted on the horizontal axis, with positive resistance to the right. The vertical dimension is reactance. Inductors have positive reactance (above the resistance axis) while capacitors have negative reactance (below the horizontal axis). The point where the axes cross is the *origin* of the plot, with value of zero for both axes.

Shown on the graph is an arrow or *vector* representing the combined resistance and reactance of a motor. The combination is called *impedance*, which must have two values (or *components*), one for resistance (horizontal value of 6 Ω) and one for reactance (4 Ω of inductive reactance).

Ideally, loads are resistive. Their impedance vectors are horizontal, with 0 Ω of reactance. Resistances are normally all positive. (A negative resistance is a kind of power source.) When they have a reactive component, this can cause some current to flow back into the source and it can also cause momentary, or *transient*, currents of high value.

Waveforms

We are usually interested in what voltage or current is doing as time goes by. A plot of voltage or current values with time, as shown below, is called a *waveform*.

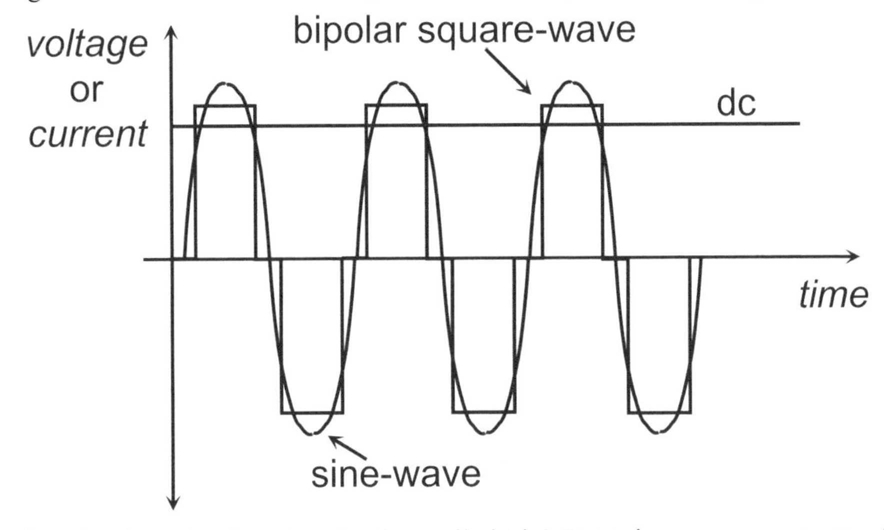

An unchanging (constant) voltage is often called 'dc'. Batteries source constant voltages, though over time they change slightly.

Inverters output changing waveforms that also change in *polarity*. The inverter waveform goes through periodic repetition. Each repetition is a *cycle*. For half of the cycle, the voltage or current is positive in polarity and for the other half it is negative. A waveform that is sometimes positive, then negative, is *bipolar* or 'ac' because it has both + and − polarities in a cycle. Some inverters output a bipolar square-wave, so-called because it looks square. Better inverters and the utility grid provide smooth sine-waves.

A cycle is one repetition of the waveform. The time a cycle takes is called the *period*, *T*. The repetition rate of cycles is the *frequency*,

$$f = 1/T$$

The cycles per second is a unit called a Hertz, abbreviated Hz. Power-line frequencies are 50 Hz (Europe) and 60 Hz (North and Central Americas). The frequency of constant (dc) waveforms is 0 Hz. The *amplitude* of a sine-wave is its peak value. These two characteristics (or *parameters*) of sine-waves, along with their waveshapes, define the waveform.

All waveforms that are not sine-waves, such as the bipolar square-wave, are composed of the sum of multiple sine-waves of different amplitudes and frequencies. A square-wave is composed of a sine-wave at the same frequency as the square-wave, called the *fundamental* frequency, plus higher-frequency sine-waves at odd integer multiples (× 3, × 5, etc.) of the fundamental frequency, called *harmonics*. These higher harmonic frequencies are usually dissipated as wasted power in motors and other electric loads. Consequently, although bipolar square-waves are easier to generate in inverters than sine-waves, they will cause some additional power loss (and heating) in loads. Also, the highest harmonic frequencies can interfere with radio and other communications. For these reasons, sine-waves are preferred to square-waves for power distribution.

Waveforms: How Much?

How much sine-wave voltage or current is there when it varies in time? We need some way of saying what the voltage or current value of a changing quantity is. One is its amplitude, or *peak value*, denoted here as $\hat{v}$ for a voltage or $\hat{i}$ for a current. When either voltage or current is meant, x, will be used to mean either. A peak quantity is designated as $\hat{x}$.

The average voltage or current value of a power sine-wave or square-wave is zero. If it is not zero, then the non-zero amount is a constant value and is the dc component added to the ac waveform. Unless dc is intentionally being distributed, as it might be from a battery bank, any dc added to an ac waveform is undesirable and can cause failures and waste power in transformers. The average value is designated here as $\overline{x}$. We are mainly interested in average power, $\overline{p}$. Average current and voltage are values that are not used much in power systems.

What *is* used much is the *rms value*, designated here as $\widetilde{x}$. 'RMS' stands for 'root mean square' and refers to the equation used to calculate it. This is the value of voltage or current that delivers the same amount of power to a resistive load as would a constant (dc) voltage or current of the same value. It is the most common value given for ac waveforms. For a 120 V power system, 120 V is the rms value. Also common is twice this voltage, or 240 V. These values can vary by as much as −15 % and −10 % depending on the utility grid. Consequently, other people use 110 V, 115 V, or 117 V as the nominal power-line voltage. I use 120 V, which is an ideal value and can be achieved in a small system such as for a home or business, but is impractical to maintain for distribution in towns or across long distances because of the voltage drop along the distribution wires.

Power

Power is calculated from *Watt's Law*:

$$p = v \cdot i$$

Besides Ohm's Law, this is the most used equation in electric power work. The convention for math symbols is that a capital V (for voltage) or I (for current) is a constant (dc) quantity. The dc form of Watt's law is that the resulting power value, $P = V \cdot I$, is the average power, $\overline{p}$. Then $\overline{p} = P$.

For changing (ac) waveforms, average power is *not* average voltage times average current: $\overline{p} \neq \overline{v} \cdot \overline{i}$! This is why average voltage and current are not used much. Instead, it is the product of rms values;

$$\overline{p} = \widetilde{v} \cdot \widetilde{i}$$

However, for peak power,

$$\hat{p} = \hat{v} \cdot \hat{i}$$

That is, peak power *is* the peak voltage times the peak current when the peaks occur at the same time.

For sine-waves,

$$\hat{x} = \sqrt{2} \cdot \widetilde{x} \approx 1.414 \cdot \widetilde{x}$$

where '$\approx$' means 'approximately equal to'. The peak of a voltage or current sine-wave is about 1.4 times greater than the rms value. What is the peak voltage (or amplitude) of a 120 V rms sine-wave? It is approximately (1.414)·120 V $\approx$ 170 V.

Knowing the relationship between peak and rms values for sine-waves, we can find the average power in peak values:

$$\overline{p} = \widetilde{v} \cdot \widetilde{i} = \frac{\hat{v}}{\sqrt{2}} \cdot \frac{\hat{i}}{\sqrt{2}} = \frac{\hat{v} \cdot \hat{i}}{2} = \frac{\hat{p}}{2}$$

Thus, average sine-wave power is half the peak power.

For a bipolar square-wave, the rms value depends on how much of the time the square-wave is 'on' or non-zero in value. The fraction of the total is called the *duty ratio* (or duty cycle), D. The range of possible values for D is from zero to one.

The rms value of a square wave is related to its amplitude (or peak value) and D by the following formula:

$$\tilde{x}(\text{square - wave}) = \sqrt{D} \cdot X$$

The amplitude of a square-wave is its constant 'on' value, and is also designated by a capital X to indicate that it is a constant.

Inverter designers must answer the following question: If $X = 170$ V, the peak voltage for a 120 V rms sine-wave, then what must be the value of D for the square-wave to also be 120 V rms? The above equation is solved algebraically for D and the resulting formula is

$$D = \left(\frac{\tilde{x}}{X}\right)^2$$

Substituting 120 V for $\tilde{x}$, 170 V for X, and then dividing and squaring, $D = 0.5$. The square-wave must be on half the time for each (positive and negative) half-cycle.

Energy

Power is the energy rate. Energy has units of joules (J) or kilowatt·hours. A joule is a watt·second. A kilowatt (kW) is 1000 watts (W). Then one kW·hour (kW·h) is 1000 W times one hour, or 3600 seconds, resulting in 3.6 million joules (3.6 MJ). Power can be expressed as

$$P = W/t$$

That is, power is energy, W over time, t. (Energy and work are the same quantity, and W is used for both.) Energy is the accumulation of power over time. If a load uses 100 W, the longer it uses it, the more energy has been used. This is similar to current as a rate of charge. If 1 A flows into a battery for an hour, the amount of charge delivered into the battery is 1 A·h. The longer 1 A is delivered, the more charge accumulates. The longer power is delivered to a load, the more energy is used by it.

Phase and Power Factor

Sine-waves of voltage and current are aligned in time, or 'in phase' with each other for circuits with resistive loads. When a load contains some reactance, such as a motor, then the voltage and current waveforms no longer align with each other but are shifted in time. This shift can also be given as a fraction of a cycle, or *phase*, which is usually measured in degrees. A full cycle of a sine-wave is divided into 360°, as circles are. A circle centered at the origin of a plot has 0° at its intersection with the horizontal axis, as shown below.

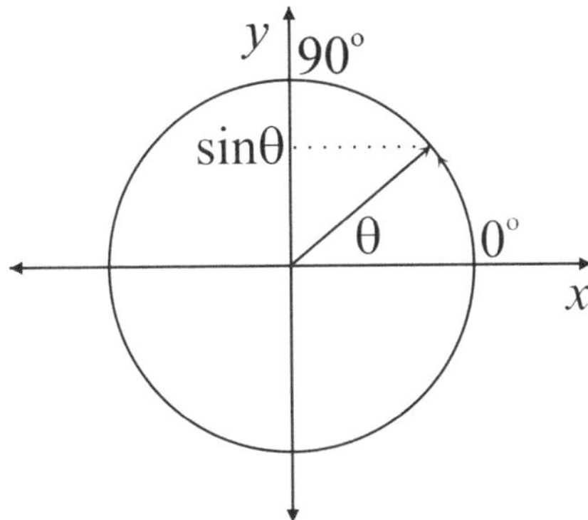

As the circle is traversed, beginning at 0° on the x-axis, the vertical value at any point on the circle is the sine of the angle, θ, abbreviated in math as $\sin\theta$ or $\sin(\theta)$, where the parentheses in math notation means that the value of the sine is dependent upon the value of the phase-angle, θ. If $\sin(\theta)$ is plotted against θ the result is a sine-wave. As a generator shaft turns, the generator windings produce a voltage that is also the sine of the shaft angle. Consequently, sine-waves and circles are related.

The circle plot also has a vector drawn with a length equal to the radius of the circle and angle from the zero-degree position of the horizontal axis. If this vector is rotated counterclockwise (CCW), its component along the vertical axis will generate a sine-wave. The frequency of the sine-wave is the number of times the vector rotates a complete revolution per second. Frequency is the rate of change of phase: 1 Hz = 360° per second.

Now suppose that there are two such vectors, one for current and one for voltage. For a load with inductive reactance, the voltage vector will lead the current vector in the direction of rotation. The difference in angle between them is important. For a pure inductive load (no resistance), the voltage vector will be at 90° (vertical) when the current vector is at 0° (horizontal). Their product is power. When it is decomposed into real (horizontal) and reactive (vertical) components of power, the formulas for each are:

$$p_{real} = v \cdot i \cdot \cos(\theta) \qquad p_{reactive} = v \cdot i \cdot \sin(\theta)$$

where θ is the angle from the current vector to the voltage vector. For a purely reactive load, no power is dissipated and $p_{real} = 0$ W. However, power is stored as energy in the inductor and is $v \cdot i$. These values can be determined from the formulas and a calculator with sin and cos functions. For the above, $\cos(90°) = 0$ (and thus $p_{real} = 0$ W) and $\sin(90°) = 1$. For a circle with a radius of one, for any point on the circle, the vertical value is $\sin(\theta)$ and the horizontal value is $\cos(\theta)$. Consequently, at 90°, the vertical value is maximum (1) and the horizontal value is 0.

In electric power systems, it is important to try to make the collective effect of all the loads appear resistive, not reactive. The measure of how well this is achieved is called the *power factor*, and is the $\cos(\theta)$ in the equation for p_{real}. Ideally, $\theta = 0°$, $\cos(\theta) = 1$, and the current and voltage waveforms are in alignment.

Solar Panels

Solar radiation is converted to electricity by a *photovoltaic* (PV) process. The conversion devices are *solar cells*. Cells are combined, as can be seen in the above picture, and interconnected as *solar panels*. The eight panels shown above form a *solar array*, mounted to a *solar tracker*. One or more arrays comprise a solar PV system.

The panels shown above have cells that are manufactured using the same processes as silicon integrated-circuit electronics. It is an equipment-intensive process with many steps and is relatively expensive. Solar cells are essentially solid-state *diodes*, electronic 'check valves' that allow current to flow one way but not the other. Solar radiation will interact with the electrons in the semiconductor materials, 'knocking' free some electrons that then become electric current. This current flows in the reverse direction than external current is allowed to flow in diodes. Each cell, when illuminated, will generate a voltage across it of about 1 V. Multiple cells are connected in series and multiple series chains are connected in parallel to form a solar panel. Typical solar panels output a nominal 12 V under full solar illumination. The sun, when directly overhead (that is, at a 90 degree elevation from the horizon) on a clear day provides a power density of about 1 kW/m^2 (1000 watts per square meter). This is considered one 'standard sun' in the PV panel industry and is the usual condition of irradiation for panel specifications. In the science of light, *irradiation* is a measure of the amount of total electromagnetic radiation (visible light + infrared + ultraviolet (UV)) whereas *illumination* refers to the amount visible to human beings.

PV panel output is not only dependent upon the amount of solar irradiation, but also upon panel temperature. As temperature increases, output decreases. Cool panels produce more power.

When a panel is loaded so that current flows from it, its voltage decreases only slightly until the 'knee' of the voltage-current curve is reached. Above the 'knee current' the voltage falls quickly, as shown in the plot below (solid curve) for a Solarex BP MSX-120 panel.

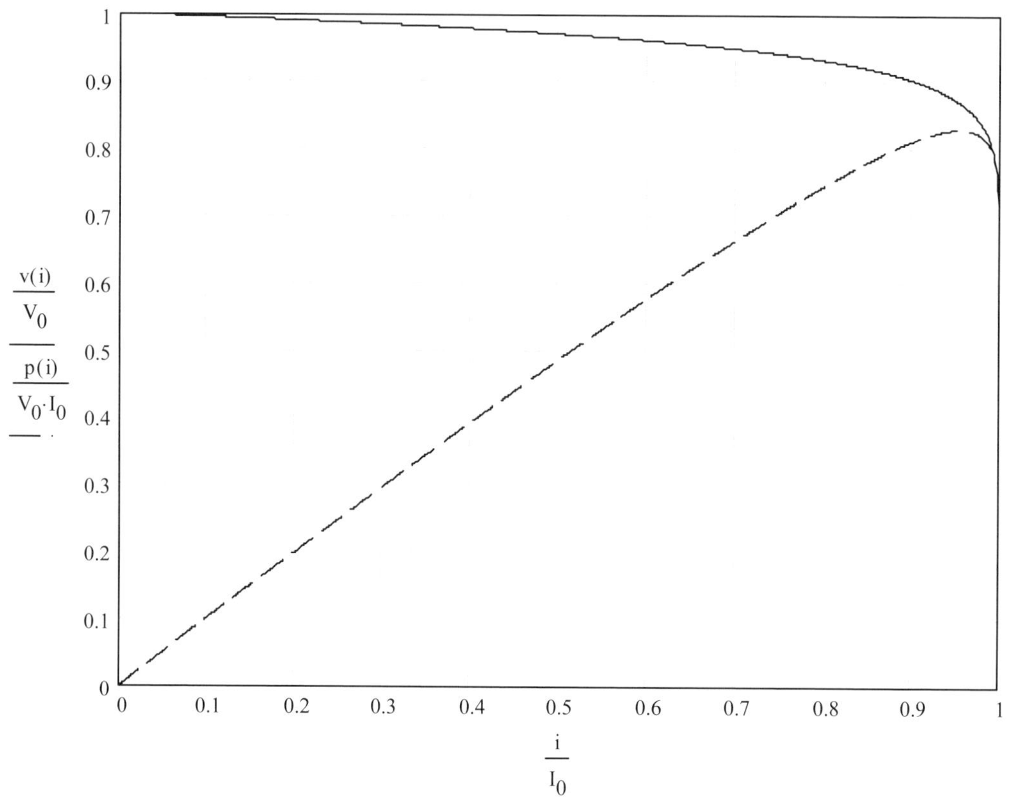

The dashed curve is that of power versus current (or $p(i)$), and it increases with current until it reaches its peak at the knee of the $v(i)$ curve. This is the optimal current, I_m, and voltage, V_m, at which to operate the panel. The point on the curve that is (I_m, V_m) is the *maximum-power point*. If the panel is operated away from this point, less than the maximum rated power, P_m, is extracted from the PV panel by the solar charger. Some solar chargers or converters are designed to operate at the panel P_m operating point on the above curves. However, not all do this, and a significant fraction of available panel power can be lost.

Trackers

Maximum power is achieved when a panel is pointed directly at the sun. A more precise way to say this is that the panel is aligned with the sun when a line perpendicular to the flat surface of the panel goes through the sun. When the angle, θ, between this perpendicular (or *normal*) line and a line from the panel to the sun is not zero, the panel is tilted from the sun, receives less radiation, and produces less power, as shown below.

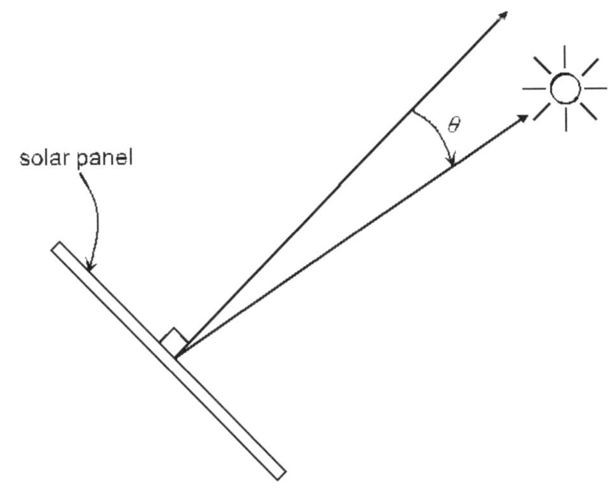

The amount of solar input power varies with the cosine of the angle. Therefore, on a clear day with an overhead sun, the panel solar input power is:

$$p_{in} = (1\ \text{kW})\cdot\cos(\theta)$$

If the panel is pointed away from the sun by 30° then the solar power the panel receives is instead:

$$(1000\ \text{W})\cdot\cos(30°) \approx (1000\ \text{W})\cdot(0.866) \approx 866\ \text{W}$$

If panels are mounted in a fixed position on a roof or stand, then only at one time during the day will they be aligned with the sun. At sunrise and sunset, θ will be largest. If the panels are mounted to point directly at the sun at noon, then during the day, θ will vary from about −90° at dawn to 0° at noon to about +90° at dusk. When the average power is calculated over a day, it is at best the average of a half-cycle of a sine-wave, which is $2/\pi \approx 0.636$. In other words, fixed-mount panels will receive only about 64 % of the available solar power, a loss of about 36 %.

Solar Panels

To avoid this loss, *solar trackers* were invented. The panels are mounted on them and they move to keep the panels aligned with the sun. Some trackers operate using electronic control. An electronic tracker can be designed using simple electronics and two light-emitting diodes (LEDs) as photosensors. Because sunlight is generally intense, LEDs, which are designed to emit light, can also be used to sense sunlight. Solar tracking is not a difficult problem. Because of the intensity of the sun, even on cloudy days, LEDs can be used as directional sensors by offsetting the angle between two of them, as shown below.

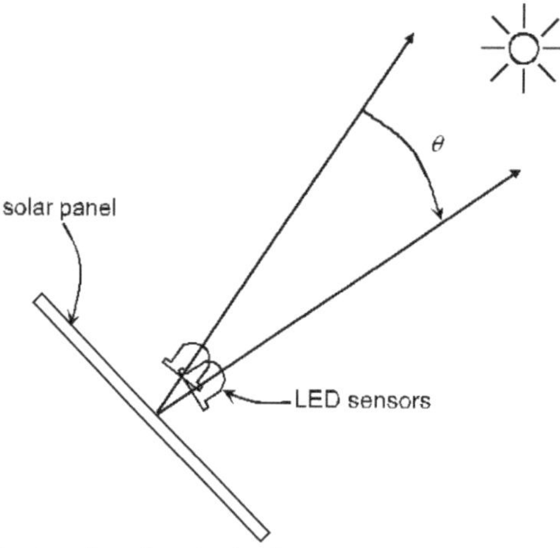

A simple analog electronic circuit (using comparators) can seek the equal-output angle, thereby tracking the sun. Additional circuitry is required to drive a step-motor that rotates the tracker. The tracker is rotated until the two LED outputs become equal, thereby aligning the panel with the sun. This solution to the problem of finding a maximum of solar input is reduced to that of the simpler problem of detecting equality of LED inputs, shown below, at 90°.

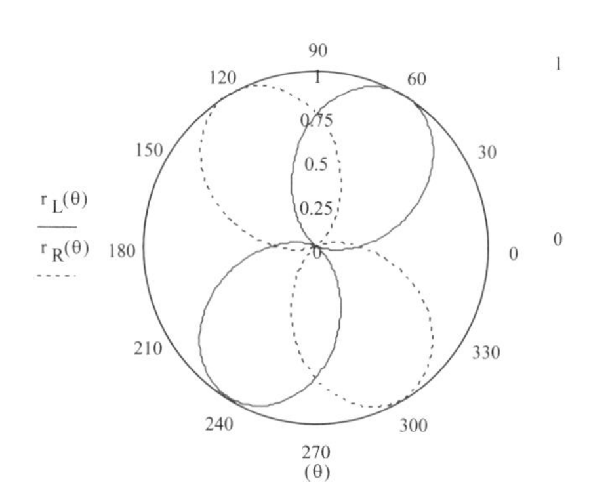

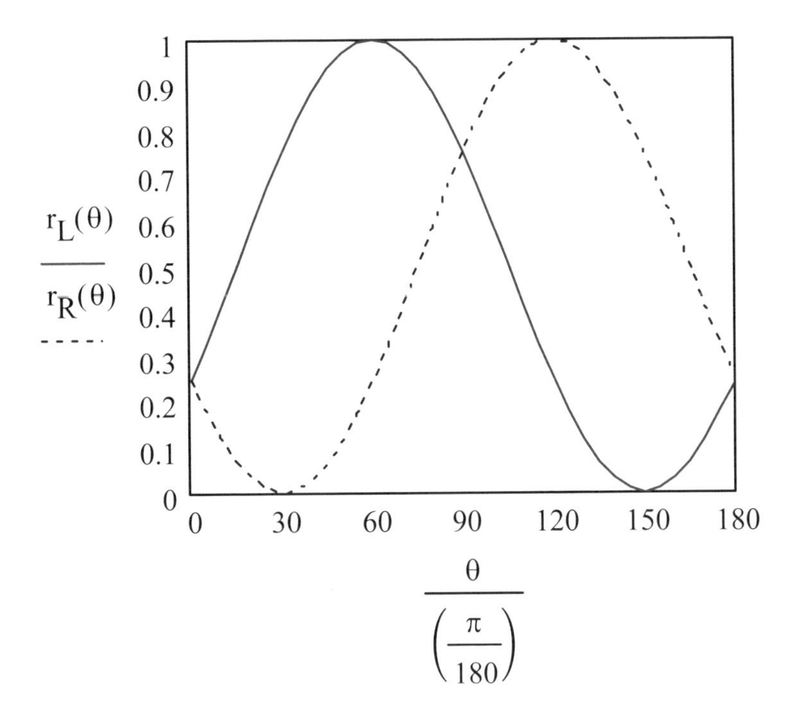

A simple and somewhat ingenious tracking scheme uses no electronics or motors (nor electric power) but instead uses the heat of the sun to rotate the tracker. Zomeworks of Arizona is a supplier. The tracker consists of a frame for holding the panels, mounted on

a central bearing, as shown in the picture at the beginning of this chapter. (The white components below the bearings are shock absorbers.) At each end of the tracker is a hollow square tube, or canister, filled with a volatile gas. The canisters are connected across the length of the tracking frame by a connecting tube. Each canister has an aluminum reflector mounted to it. When the sun shines onto the inside of the reflector, as shown, the canister is heated and its vaporized gas flows through the connecting tube to the cooler canister. The reflector of the cooler canister reflects solar radiation away from its end of the tracker, thus keeping the canister relatively cool. The weight of the additional gas causes the tracker to rotate in the direction of the cooler canister. Another view, from above two trackers, is shown below.

The disadvantage of thermal over electronic trackers is that they are slow to 'wake up' in the morning. The sun's rays are weaker, attenuated at both early morning and at dusk by the atmosphere. The longer path in air weakens, or attenuates, the radiation by absorption and reflection. At both the beginning and ending of daylight, the solar intensity is much reduced, and so is the heating effect on the canisters.

The problem of losing early solar power is mitigated by using a couple of sticks or plastic PVC plumbing tubes to prop up the trackers in the evening, so that they are pointing toward the sunrise. Leave some range for the trackers to lift up off the props once they begin to track so that the props will fall away, leaving the trackers free to rotate with the sun in the opposite direction later in the day.

The trackers shown in this chapter cost about $1000 US (2003) each. Whether their additional cost is justified depends on an acceptable payback period. If solar panels decrease in price, then eventually, fixed-mount panels will be justifiable. However, at $5 US per watt for panels, the additional one-third output the trackers provide break even around 600 W of panels. For smaller and lower-priced trackers which hold lower-power panels, the tradeoff is similar, but with a scaled-down wattage.

Panel Wiring

Panels are essentially voltage sources and have two terminals. The electrical connections of a panel are found in a junction box on the backside of the panel, as shown below.

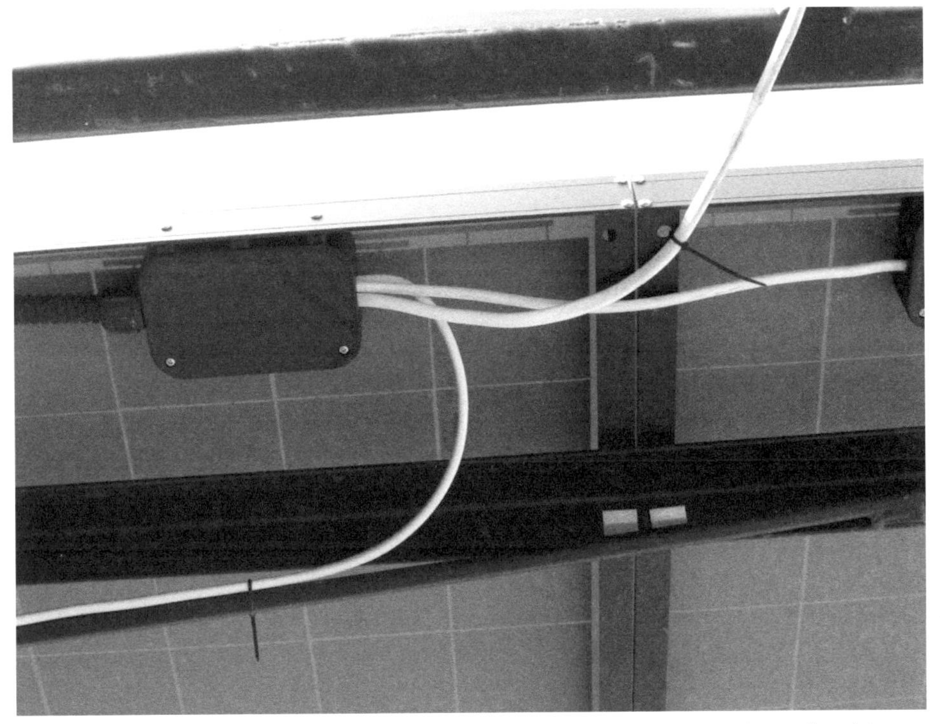

This panel is represented in the manufacturer's information sheet (in this case, a Solarex BP MSX-120) as shown below.

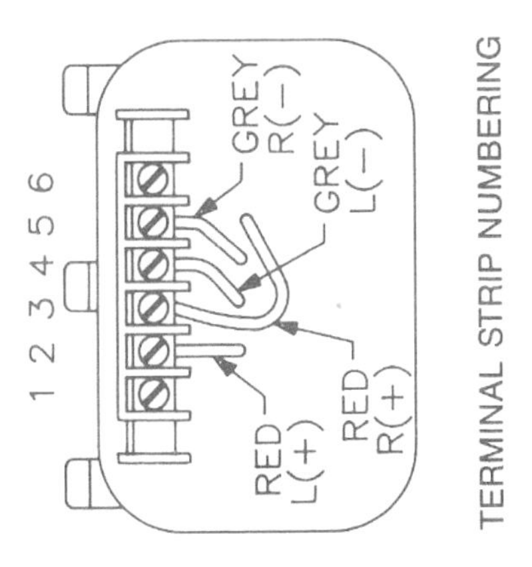

Solar Panels

For this panel, two separate solar-cell circuits are brought to the terminal strip, each with a positive (+) and negative (−) terminal. It is as though the single panel were two panels joined into a single unit. The terminal strip inside the junction box has screw-down connectors for holding wires placed underneath the screws. Connectors numbered 2 and 3 are the positive terminals and 4 and 5 are negative. (This drawing does not show that terminals 2 and 4 are one source and 3 and 5 are the other.)

Multiple panels as PV sources must be combined in some configuration which complicates wiring somewhat. The cells in panels are connected in series, but multiple strings of series cells are usually connected together in parallel (or shunt). Often, larger panels such as this one are divided into two series strings of cells with 6 V outputs each. Then if both series circuits are brought out to the junction box on the backside of the panel, there are really two voltage sources and two pairs of terminals, as show in the circuit diagrams below.

6 V (parallel) connection

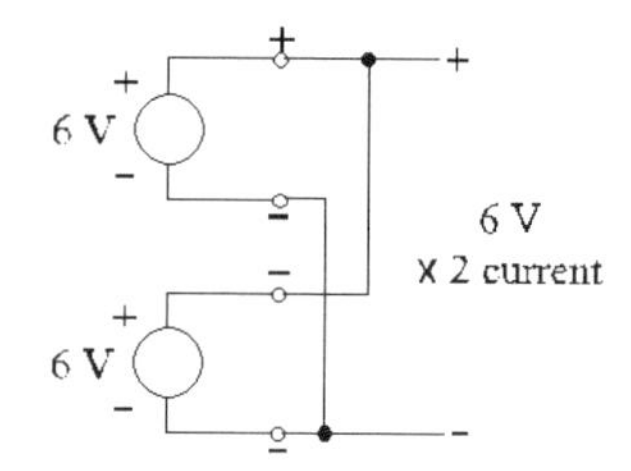

12 V (series) connection

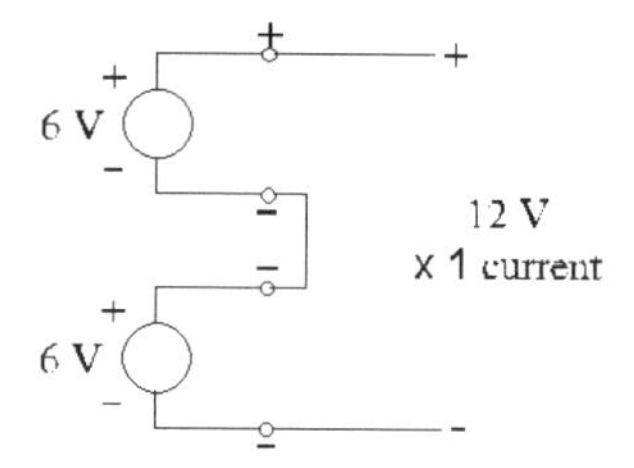

The information sheet accompanying the panel presents this information in the following drawings of the junction box and its connections.

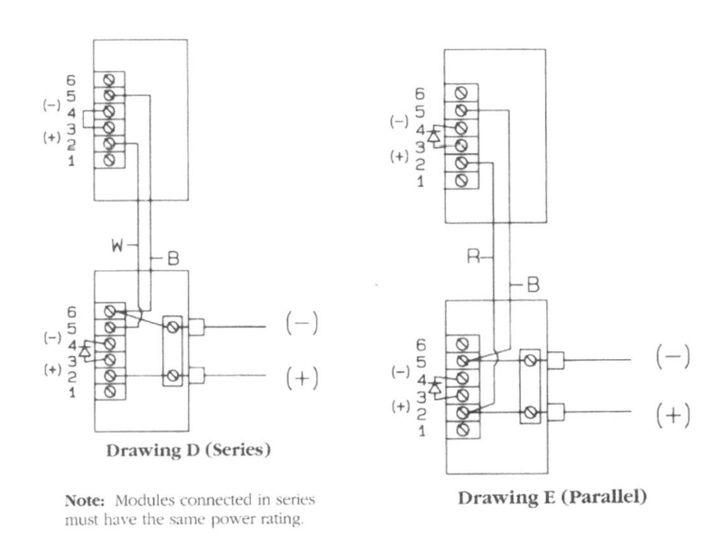

Drawing D (Series)

Note: Modules connected in series must have the same power rating.

Drawing E (Parallel)

The circuit of drawing D (left) shows how two junction boxes can be connected with sources in series. The upper box is wired equivalent to the previous series circuit diagram. The terminal numbers correspond as shown below.

The circuit diagram shows an additional component that looks like this.

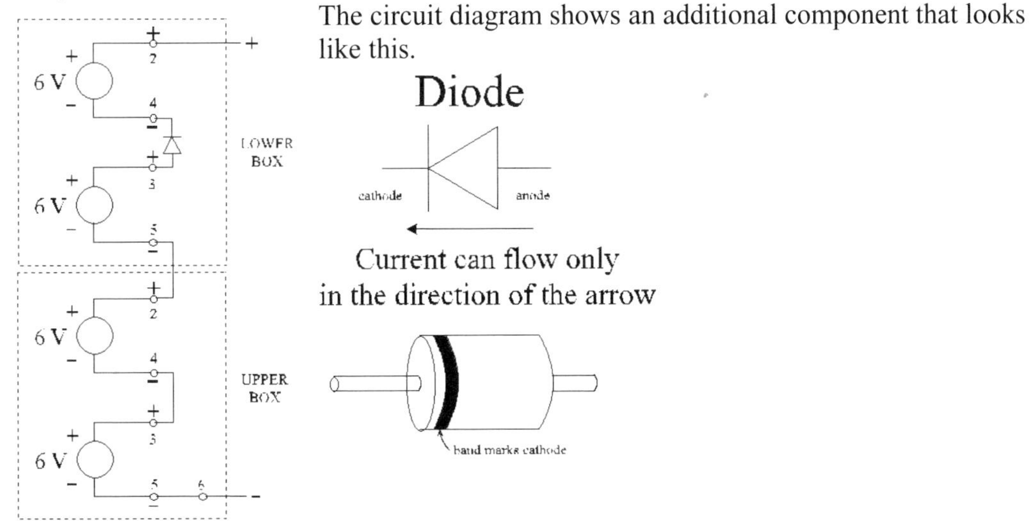

Solar Panels

A diode functions like a check valve in a water system. It will allow current to flow only in the direction of the arrow in its symbol. Diodes are added to solar panel wiring because of partial panel shading. If some of the panel is in shade and not producing current, the diodes bypass it, allowing current from illuminated cells to flow to the output instead of into the shaded cells. Each parallel string of cells contains a diode somewhere within the series string. In the above circuit, it is placed between terminals 3 and 4.

In the above circuit, each of the two junction boxes was wired for a series output of 12 V. Then the two panels were also wired in series, for a final output of 24 V. The output can then be wired in parallel with other groups of panels of equal current rating. The resulting voltage remains 24 V but the currents add (by Kirchhoff's Current Law) from each parallel branch. The branches should have equal current capability so that no one branch dominates and tends to suppress the current output from the others. Connecting cells in parallel is a kind of balancing act. When part of a panel is shaded, the shaded branch of cells will have less voltage output and the series diode will simply disconnect them from the other branches so that current from the other branches does not flow into the shaded branch in the reverse direction.

The diodes have a voltage drop across them when conducting of about 1 V. Using Watt's Law (p. 11), this causes a power loss in the diodes of the conducted current times 1 V. The loss is undesirable but also unavoidable in a simple system. (It is possible to provide a solar charger for each individual branch of the solar array to avoid the diodes but with the present technology would be very expensive.)

The manufacturer's Drawing E (above) wires the cells within the junction box in series, for 12 V, but then wires the two junction boxes in parallel. The output is 12 V but with twice the current capability. A lower voltage, higher current configuration will require larger connecting wire to the solar charger. The series configuration (24 V, half the current) outputs the same amount of power but at four times the source resistance. ($R = V/I$; then $2 \cdot V/(0.5 \cdot I) = 4 \cdot R$.) House wiring insulation is rated for over 200 V, and by raising the solar-array output voltage, smaller (and less expensive) wire can be used without disadvantage. However, not all solar chargers operate optimally with a higher voltage. This is a system design issue.

When the junction box of the MSX-120 is opened, the interior wiring is revealed.

The looped red and gray wires come out from the sealed panel to attach to the terminals. The larger red (+) and black (−) wires above them are the output wires leaving the box at the top. On the terminal strip, the wiring corresponds to the following circuit.

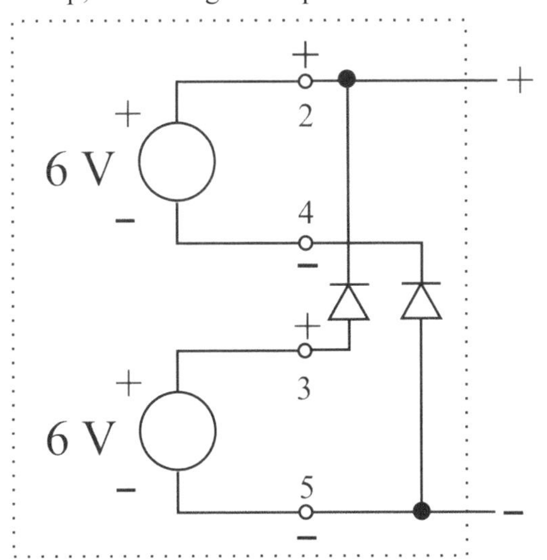

The two diodes are placed in series with each 6 V cell string because these strings, or circuit branches, are then connected in parallel. (For the MSX-120, the sources are actually 12 V instead of the more usual 6 V.) Note in the photo that the bands marking the cathodes of the diodes are silver and the diode body is black. Markings vary, and might even be a '+' sign instead of a band.

The junction box corresponding to the connection of two boxes in parallel is shown below. Usually this results in a 6 V array; for the MSX-120, it results in a 12 V configuration. Then 12 V panels wired in series pairs result in a 24 V array output voltage.

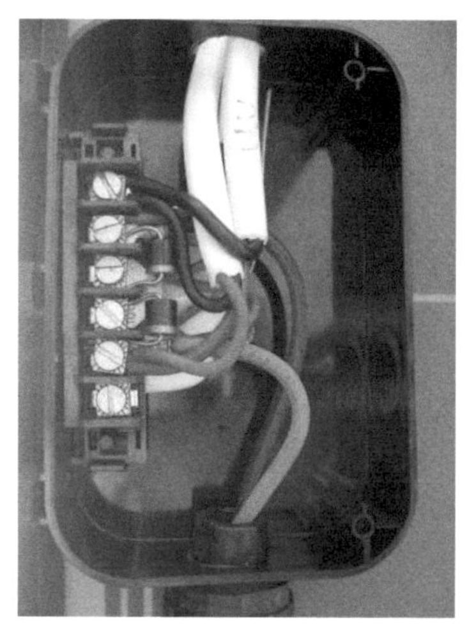

In mild climates, panels can be connected with house wiring without placing the wiring in conduits. This is shown below for a solar array wired in the subtropics.

However, from the array, mounted on either a moving tracker, roof, or stand, the wiring to the rest of the system should be more protected, being run in either 20 mm or 25 mm plastic electrical tubing (*conduit*). For underground passage, PVC plumbing pipe works well and can be constructed in the same way as plumbing. An example is shown below.

The connection between array wiring and conduit wiring can be a continuous cable. However, it is convenient for system installation and diagnosis of problems to be able to access the array outputs. This can be done simply by using a wire nut connection for electrical access, as shown below. In harsher climates, use an enclosure.

For a large array, multiple cables (two shown here) not only reduce power loss in the wiring but also provide flexibility in the energy center. By placing different halves or quadrants of the array on separate circuit breakers in the PV array panel, they can be switched in and out separately when searching for electrical faults or measuring the fraction of current from each part of the array.

Battery Banks

Batteries are chemical to electrical power converters and are power sources when charged. When they are charging, they become electrical to chemical converters. Because batteries in use must be continually recharged, they function as charge storage reservoirs. A reversible electrochemical reaction proceeds in one direction while discharging and in the opposite direction while charging.

A *battery bank* is multiple batteries connected to function as one large battery. Usually these batteries, as the ones shown above, are 12 V lead-acid batteries. These batteries are made differently than the ones typically found in vehicles, which are *starting batteries*. They are never discharged very much and are optimized to provide maximum starting current, as rated in 'cold cranking ampères'.

The batteries used in home electric systems are *deep-cycle batteries*. The best and costliest are Plante batteries. Deep-cycle batteries have thicker lead plates and are capable of lasting for more charge-discharge cycles when discharged more extensively or 'deeply'. (The unconnected battery shown above is a starting battery. It looks like a deep-cycle battery from the outside. The deep-cycle batteries are labeled as such.) Battery life is reduced at increased temperatures but is most affected by the extent and duration of discharge. The quickest way to wear out batteries is to leave them in a discharged state. A fully charged battery will last for years, until the plates warp or sulfate, cells short, or sulfur in the sulfuric acid electrolyte precipitates out of solution.

Battery Charge Capacity

The ability of a battery to store electric charge, Q, is rated in ampère·hours, or 'amp·hours', A·h. (Another unit of charge is a *coulomb*, C, which is an ampère·second or 1/3600 of an amp-hour.)

As batteries age, the amount of charge they can store - their charge 'capacity' - decreases until the storage charge is unacceptable. This is usually at a value of about 50 % of the rated charge capacity.

The charge capacity is not a fixed number. It might seem as though a battery of a certain volume should be able to hold a certain fixed amount of electric charge. However, the effective available charge depends on multiple factors. One is the discharge rate. The charge capacity rating is usually given by battery manufacturers for a discharge rate of Q/20 h. A 220 A·h battery discharging at 11 A will be fully discharged after 20 h. As the discharge rate (which is current) from the battery is decreased, the total charge capacity is increased.

Battery-Bank Voltage

Like solar panels, batteries can be wired in series to increase the battery-bank voltage. The higher the voltage, the less current is required for the same power. Doubling the voltage from 12 V to 24 V halves the current. However, inverters and solar chargers have specified voltages and these will constrain the choice of battery-bank voltage to maintain compatibility with electronics components of the system.

Lead-Acid Batteries

Lead-acid batteries are of three major types:

- Wet, flooded, vented, or unsealed: these are the conventional open batteries that have been around for a long time. They must be handled carefully to avoid spilling electrolyte from them, though the electrolyte can also be checked by removing cell caps and replenished when needed.
- Gelled electrolyte or gel-cell: the electrolyte is in a gelled form, the battery is sealed, and oxygen and hydrogen are recombined through small fissures or cracks in the gel. The battery can be charged more quickly than a wet cell and can have a longer life in hot weather but can also be ruined through overcharging more easily. It can be operated in various orientations.
- Absorbed glass matte (AGM): also sealed, these have a liquid electrolyte that is held in contact with the lead plates with a matte. The recombination of oxygen and hydrogen also occurs within the closed volume of the cells.

Both gel-cell and AGM batteries are also referred to as 'valve regulated' or *sealed-lead-acid* (SLA) batteries.

Battery Electrochemistry

It is not necessary to understand battery chemistry to use batteries or to design an electric system. Some insight into what happens internally, however, can aid in understanding battery behavior and symptoms of battery problems.

Each terminal of a lead-acid battery has a different electrochemical reaction. It is *electro*-chemical because it involves the flow of electric charge (as electrons) that participates in the reactions.

At the positive (red) terminal (or *electrode*) of the battery, electrons (which are negatively charged) flow into the terminal when discharging. The chemical reaction for discharging proceeds from left to right as given below, and for charging, from right to left:

$$PbO_2 + SO_4^{2-} + 4H^+ + 2e^- \leftrightarrow PbSO_4 + 2H_2O$$

During discharge, lead sulfate ($PbSO_4$) is deposited on the plates. The charged plates are covered with lead dioxide (PbO_2) and the electrolyte consists of sulfuric acid, H_2SO_4, and water. However, in solution the acid does not remain as a molecule but decomposes into *ions* - charged parts of molecules - which are a positively-charged hydrogen (H^+) and a negatively charged sulfate grouping or *radical*, SO_4^{2-}, plus two free electrons (e^-). The lead bonds with oxygen of the water forming PbO_2 and leaving the hydrogen as positive ions in solution. When electrons (negative charge) enter the positive terminal, this is the equivalent of positive current flowing *out of* the positive terminal, as we have assumed by convention in circuit analysis. These entering electrons add to the reaction equation and cause oxygen to be released from the lead which then bonds to the sulfate radical. The oxygen recombines with the hydrogen to form water. During discharge, the water level in the battery rises.

For chemical equations, both charge and number of atoms of each element must be the same on each side of the equation. Because of mass balance, the amounts of each element are the same because matter is neither created nor destroyed, nor is it *transmuted* from one element to another.

At the negative (black) electrode, the reaction is

$$Pb + SO_4^{2-} \leftrightarrow PbSO_4 + 2e^-$$

Elemental lead (Pb) and a sulfate radical in solution combine to free two electrons to leave the negative terminal. Two entered the positive terminal, and the amount of current is the same for each terminal (by KCL; current in a closed loop is the same everywhere in the loop). During discharge (left to right in the reaction equations), lead sulfate also forms on the negative plate.

In wet-cell batteries, another pair of reactions are also occurring. When discharging, oxygen at the positive electrode combines with hydrogen ions and electrons to form water:

$$\tfrac{1}{2}O_2 + 2H^+ + 2e^- \leftrightarrow H_2O$$

At the negative electrode, hydrogen decomposes into hydrogen ions and free electrons:

$$H_2 \leftrightarrow 2H^+ + 2e^-$$

In SLAs, the decomposition of water into hydrogen and oxygen is reversed during charging (right to left reaction) by diffusion of oxygen from the positive electrode through the immobilized electrolyte where it combines at the negative electrode with hydrogen ions to form water. This occurs either electrochemically as

$$H_2O \leftrightarrow \tfrac{1}{2}O_2 + 2H^+ + 2e^-$$

or chemically as

$$H_2O + PbSO_4 \leftrightarrow Pb + \tfrac{1}{2}O_2 + 2H^+ + SO_4^{2-}$$

For SLAs, water is replenished by these charging reactions. Loss of water limits the life of SLAs.

Electrolyte

Lead-acid electrolyte is sulfuric acid solution at a concentration of 35% sulfuric acid and 65% water. The concentration can be measured indirectly with a hydrometer, a low-cost instrument that measures the density, ρ, of a liquid relative to the density of water, ρ_{H_2O}. This is the *relative density* or 'specific gravity' of the solution;

$$\text{specific gravity} = \frac{\rho}{\rho_{H_2O}} = \frac{\rho}{1\,\text{kg/l}} = \frac{\rho}{1\,\text{g/ml}}$$

at 25 °C (77 °F). A litre, abbreviated l, is a thousandth of a cubic meter, m^3, and a milliliter, ml, is the same as a cubic centimeter, cm^3 ($1\ ml = 1\ cm^3$). A US gallon is about 3.785 l, making a litre somewhat larger than a quart. Specific gravity is a ratio of densities of the same units and has no unit.

Hydrometer readings must be corrected with respect to temperature so that they are what would be expected at 80°F (27 °C) for use of the following table. The correction is based on a change in hydrometer reading of 0.0004 per °F. Measure the temperature and subtract 80 °F from it. If the temperature is less than 80 °F, the result will be a negative number. Then multiply it by 0.0004. Add the result to the hydrometer reading. (If the result is negative, it is subtracted as the addition of a negative number.) A fully charged battery should have a reading of somewhere between 1.270 and 1.284, or 1.277 ±0.007.

Hydrometer readings should not vary between cells more than 0.05. If the specific gravity is low after fully charging, then an equalization cycle should be run and the specific gravity checked again. If the value is still too low, try desulfation. (See below.)

The battery voltage as given in the table is an easier method for determining battery charge. However, it is only valid for a battery that has neither been loaded (is open circuited) and that has also not been charged for a few hours. This is not possible in an operating system, and voltage can only be used to estimate battery charge. For a more precise determination of charge, use the hydrometer.

Charge, %	Specific Gravity @ 80°F	Voltage, V
100	1.277	12.73
90	1.258	12.62
80	1.238	12.50
70	1.217	12.37
60	1.195	12.24
50	1.172	12.10
40	1.148	11.96
30	1.124	11.81
20	1.098	11.66

Source: *Deep Cycle Battery Maintenance*, Trojan Battery Company

For cold-climate battery use, the freezing temperature of the electrolyte varies significantly with temperature. A fully charged battery with a specific gravity of 1.28 freezes at −92 °F while at 20 % charge (1.10), at -19 °F, and at 40 % charge (1.15), at +5 °F. Battery charge capacity increases with temperature; in hot climates, batteries have somewhat greater capacity.

Battery Life

One question about battery-bank size is whether it might not be better to have multiple banks instead of one. The answer is that it is not; one bank will maximize battery life through minimization of the depth of discharge. If batteries are discharged only a small amount (discharge is 'shallow'), the number of charge-discharge cycles that the batteries can provide is increased and their life is lengthened. Ideally, the battery-bank should be as large as possible within budget and facility constraints. A huge battery bank costs a huge amount but lasts a huge amount of time too because it is hardly discharged. A small bank costs less, takes up less space, and must be replaced more often.

An approximate estimation of battery life is found from industry tests. A simplified graph of battery life versus amount of discharge is shown below for lead-acid wet cells.

Battery Banks

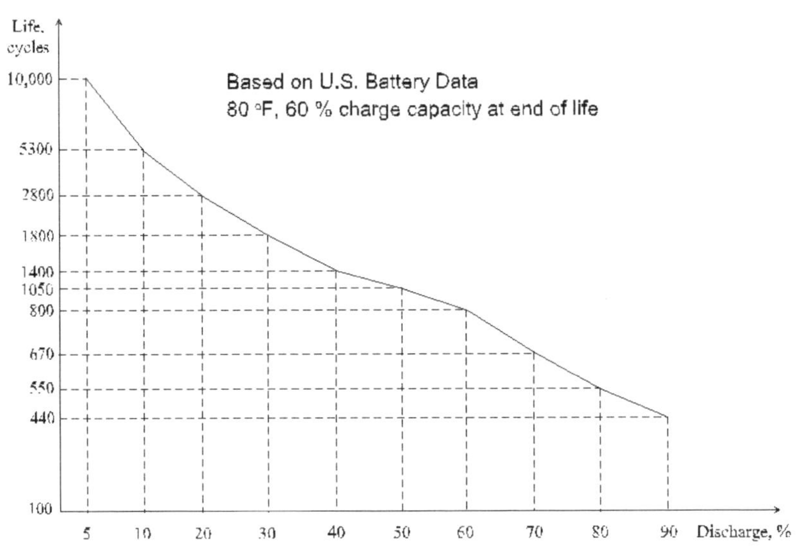

A daily charge-discharge cycle of 60 % of available battery charge results in a battery life of about 2.2 years at a nominal 80 °F (27 °C). By doubling the battery bank charge capacity, discharge is reduced to 30 % for the same loading and the battery life is slightly more than doubled, to about 1800, a 2.25 times increase in life. With a battery bank four times the size, life increases by 5.06 times. The larger bank has larger than proportional life extension. This suggests that it is best to have one bank, made as large as possible, for maximum battery life.

Maximization of the number of discharge cycles, however, is not the only approach to assessing battery life. A battery bank will not deliver as many ampère-hours of charge over an extended number of cycles as a bank discharged more deeply and lasting for fewer cycles. By multiplying the number of cycles by the fraction of discharge from the above graph, and then multiplying this number by the battery-bank charge capacity, the total charge stored and delivered by the battery over its life is calculated. For a 220 A·h battery discharged 5 %, the amount of charge stored per cycle is

$$(220\ \text{A}\cdot\text{h})\cdot(0.05) = 11\ \text{A}\cdot\text{h}$$

At this amount of discharge, the battery will last for 10,000 cycles. The total charge thus stored by it over its life is 110,000 A·h.

The same battery discharged 50 % will last 1050 cycles, with a charge per cycle of (220 A·h)·(0.5) = 110 A·h. Then for 1050 cycles, the total charge is 115,500 A·h. Though its cycle life is shorter, the more deeply discharged battery provides more charge storage over its shorter life. Additionally, more deeply discharged batteries can be recharged more quickly, reducing generator run time.

New batteries, like milk, should be bought with the most recent manufacturing date possible. Batteries usually have a manufacturing date code. The month is indicated by a letter beginning with A for January and skipping the letter I. (It looks too much like 1.) The following number indicates the last number of the year. For instance, '7' indicates 2007. D7 indicates that the battery was manufactured in April, 2007.

Charging Rate

The rate at which batteries can be charged differs with battery type and discharge. A more discharged battery can be charged at a higher charge rate or current. A high charge rate is desired if the electric source is a generator, to minimize generator run time and operate it at maximum efficiency, closer to its rated power.

According to U.S. Battery, at 20 % to 50 % discharge, wet lead-acid batteries have an acceptance rate of about 25 % of their total 20 hour amp-hr rating. Stated another way, a lead-acid battery bank consisting of three 12 volt batteries at 220 A·h each (660 A·h total) would have a charge acceptance rate of 165 amps. Then lead-acid wet cells can be charged at a rate of up to about

$$I_{\text{charge}} = \dot{Q}(\text{lead-acid wet cells}) = \frac{Q}{4\ \text{h}}$$

where the dot over the Q indicates that it is the *rate* of charge, or current. For example, a 220 A·h wet-cell battery can be charged at a maximum rate of 220 A·h/4 h = 55 A. For lead-acid gelled electrolyte batteries, the typical maximum charging rate is

$$I_{\text{charge}} = \dot{Q}(\text{lead-acid gel cells}) = \frac{Q}{2.5\ \text{h}}$$

and for absorbed glass matte (AGM) batteries,

$$I_{\text{charge}} = \dot{Q}(\text{lead-acid AGM cells}) = \frac{Q}{1\ \text{h}}$$

For a 220 A·h gel cell, maximum charging rate is 88 A and for the AGM battery, it is 220 A. The gel-cell and AGM batteries are more expensive than the older unsealed or vented wet lead-acid batteries, but the shorter charging times can reduce generator running time which also saves in operating cost.

Sulfation

The dominant cause of battery failure is *sulfation*. What makes batteries decrease in their ability to store charge is sulfur. The sulfuric acid of the battery decomposes and combines with lead of the plates to form lead sulfate ($PbSO_4$) plus water ($2H_2O$). The coated area of plates no longer participates in the electrochemical activity of the battery, effectively shrinking the size of the battery.

The most common cause of sulfation is not fully charging the battery or leaving it for some time not fully charged. In hot weather, even a day between charges can result in significant sulfation. Low electrolyte levels exposing plates to air will cause the exposed area to sulfate.

As sulfate binds to the lead, it comes out of the electrolyte solution causing it to be lower in sulfuric acid concentration. Sulfation begins when the specific gravity of the electrolyte falls below 1.225 or battery voltage measures less than 12.4 V when fully charged.

The function of reviving 'dead' (sulfated) batteries is *desulfation*. Some chargers are equipped with this function. The battery is driven with high-frequency current pulses, to decompose the crystallized lead sulfate and return the plate to either lead or lead dioxide. Battery engineering still has many unanswered questions and one is how to extend battery life by defeating life-limiting mechanisms such as sulfation.

Battery-Bank Wiring

Batteries are low-resistance (low voltage, high current) devices and the wire used to connect them is large - typically # 6 AWG to # 2 AWG for home electric systems. Reference to the wire table on page 7 shows why this range is common.

Battery connections are usually made to 'posts' with clamping connectors that fit around them, as shown below.

These connectors use bolts to tighten the connections to both battery post and wires. No more than about three # 6 AWG wires can be connected to a single post connector (as shown in the positive terminal connection) and still maintain a low-resistance electrical contact. As shown in the opening picture of the chapter, battery-bank wiring should be kept as short as possible and in the open where it can be cooled by the air.

When stripping insulation from wires, avoid cutting strands of wire, thereby reducing the current capacity ('ampacity') of the wire and possibly creating a high-resistance connection and ohmic heating (I^2R). Copper has a positive temperature coefficient of resistance; the hotter it becomes, the higher its resistance, causing even more ohmic heating and a possible runaway meltdown. After the end of a wire is stripped of insulation, twist the strands together to break any copper oxide coating on them and to bind them together so that no loose strands are left out of the connection. Use color coding while wiring: red = + and black = −.

Battery Alternatives

The dominant storage technology for home energy systems is presently lead-acid, wet-cell, deep-cycle batteries. While they provide a needed function, they also require periodic maintenance and have limited cycle life. One alternative is to store heat. (See the chapter titled 'Solar Thermoelectric Technology' in *Eco-Electrical Home Power Electronics* for details.) However, for solar photovoltaic or wind power sources, electric charge storage is the direct approach.

One possibility is to use a capacitor bank for charge storage. Capacitors have an indefinite cycle life but are limited in charge density. The highest-density capacitors have been electrolytic types. To replace a battery bank with them would, however, be prohibitively expensive and require a large space.

In recent years, the double-layer or 'electrochemical' capacitor has emerged as a replacement technology for batteries. 'Electrochemical' is misleading in that they do not involve chemical reactions, as do batteries. Sometimes they are referred to by the nondescript word 'ultracapacitors'. These capacitors are a variant on electrolytic technology, yet have much greater charge storage density. Present applications include UPSs and electric automobiles.

One leading supplier is Maxwell Technologies (www.maxwell.com). Their MC Energy series of capacitors is claimed to have a life of over a million charge-discharge cycles. In home-power use, with one cycle per day, they would last 2738 years. Obviously then, the major barrier for their use is the one-time purchase cost. The Maxwell BCAP3000-P270-T04 Electric Double Layer Capacitor (EDLC), has a length of 138 mm and OD of 60 mm, operable over a temperature range of -40°C to +65°C. One supplier offers them for $116 US each in quantities of 10 to 49. They have a capacitance of 3000 Farads ± 20% and a voltage rating of 2.7V. For a 12 V system, a series stack of 4 capacitors results in a 10.8 V bus - rather low. With 5 in series, the capacitor-bank voltage is 13.5 V. Stacking equal-value capacitors in series decreases their capacitance by their number. A stack of 5 thus has a capacitance of (3000 F)/5, or 600 F. A farad (F) is a coulomb (C), or ampère·second (A·s) per volt. (1 A·h = 3600 C) The charge storage per capacitor is 8.10 kC.

A bank of 12 220 A·h batteries operated to 50 % discharge has an effective charge storage capacity of 4.752 MC, or 4752 kC. Therefore the number of these 3000 F capacitors required to replace it is

$$\frac{4752\,\mathrm{kC}}{8.10\,\mathrm{kC}} = 587$$

At $116 US each, the total replacement cost is about $68,000 US. As one's hopes sink in view of this number, another price-related quantity, the rate of decrease, raises them again. Over the '00s decade, the per-coulomb price has decreased by about 100 times, and as production volume increases, it should continue. Partial replacement of battery banks with double-layer capacitors is a present possibility in that capacitors can be added to battery banks. If prices continue to fall, individual batteries can be replaced over time as your budget allows. The earlier, more costly capacitors benefit you longer, thus partially recouping their higher price. Eventually, the battery bank becomes wholly a capacitor bank.

Solar Chargers and Converters

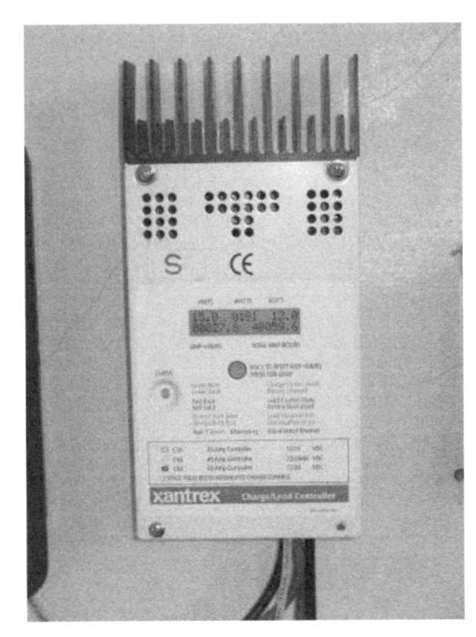

A *solar charger* inputs power from a solar array and outputs power to a battery bank. It is a battery charger operated from solar panels. It is one of two electronic blocks of the electric system shown in chapter one. Solar chargers are specified by their maximum charging current, battery voltage, whether they operate at the maximum input power, and features such as an instrument front-panel, as on the Xantrex C-60 charger shown above. They are often (though not necessarily) made to be wall-mounted.

Another system approach is to use a *solar converter* to convert from the solar array input to a *system bus* voltage of 170 V dc. This is the peak voltage of a 120 V rms ac sine-wave and the dc voltage needed by an inverter. It also can power most commercial battery chargers. This 170 V HVDC system is described in more detail in the chapter, 'Power System Design' of this book and also in 'Off-Grid System Strategy' of *Eco-Electrical Home Power Electronics*. It has advantages over the battery-bus system, as described previously in 'The Typical Home Electric Power System'. Chargers such as the Xantrex charger shown above are designed for the conventional system. They consist of two functions, as shown below.

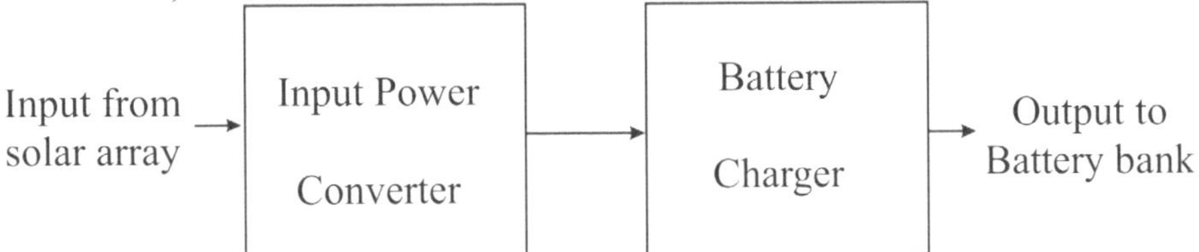

The input-power converter block converts the variable-voltage array output to a voltage that the battery charger can use to make a controlled charge of the battery bank. In some simpler designs, such as the Xantrex C-60, the two are not separable. The input converter in better designs controls the amount of current from the array and operates it at its maximum power point. In microcontroller (μC) based charger designs, a search for the maximum power point is done in software. Array output power follows the curve previously given for arrays and is copied below.

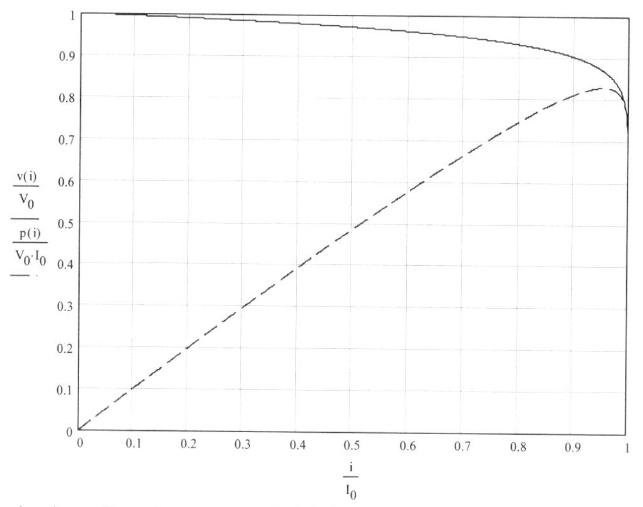

The μC will begin by allowing some initial current from the panel. Suppose (on the graph) it is 80%, or 0.8, of I_0, the short-circuit current. This point corresponds (on the vertical axis of the graph) to a power of $0.75 \cdot P_0$, where

$$P_0 = V_0 \cdot I_0$$

and is the maximum theoretical power corresponding to the upper-right corner of the graph (at (1, 1)). An ideal panel would include this point in its curve and be optimally operated there. Instead, actual panel *V-I* curves decrease in voltage with increasing current. As voltage remains fairly constant and current increases, power increases linearly, and the power curve below a fractional current of about 0.8 is fairly linear. However, as the voltage curve begins to fall off, the power curve peaks and also falls off to zero at I_0, which is the short-circuit ($V = 0$ V) current. The peak-power point corresponds on the *V-I* curve to a voltage of about $0.84 \cdot V_0$ and $0.95 \cdot I_0$. A *maximum-input-power* solar charger or converter, also called a 'maximum power-point tracker' in some commercial literature, scans through the range of current values around the maximum-power value until it finds it, then operates there. The maximum power can be found by sensing both input current and voltage, converting them to digital form as input to the μC, and then multiplying them in the max-*P* software routine.

One method of doing this search is to start with extreme or endpoint values for *i* and measure *P* at them, then choose a point midway between the endpoints and measure it. Make the new endpoints those of the *i* values corresponding to the two larger values of power, find the midpoint between them and iterate again. Do this until the two points are sufficiently close. They will be at the max-*P* value of *i*. If this search routine is run every so often, the max-*P* point will be tracked as the panel characteristics change due to temperature and illumination.

A simpler and slightly less optimal method is to maximize the battery current. If the battery voltage is roughly constant, then maximum current corresponds to maximum power. Battery current and voltage are both measured in a good charger design, and a maximization routine can be implemented by simply maximizing the output current to the battery.

Simpler solar chargers that do not track max-*P* can be designed to be fairly close, based on an assumed solar-array voltage. However, beware that such solar chargers are sometimes specified to operate over a wide input-voltage range. If the unit is designed to operate close to max-*P* with a 12 V array, but your array is wired for 24 V instead (to reduce wiring), then you might get only half the achievable power from the array with a non-max-*P* charger.

Charger Wiring

The Xantrex C-60 will be used here to illustrate typical solar charger wiring. Other units will have a different physical appearance, but the terminals and readout functions should be similar.

Four connections are made to the charger: two input and two output. The two input connections are from the solar array and the other pair go to the battery. Both are dc and voltage polarities must be carefully observed, for many chargers have no protection against reversed connections to sources. Both battery and array are sources, though the battery is the load - the recipient of power - under normal operation. When the sun goes down, however, the charger maintains its low-power functions from battery power and electronics on the circuit-board is powered by the continuous power from the battery.

Shown below is an opened C-60 charger with major features pointed out. When the unit is mounted, the heat sink should point upward, as shown, to efficiently convect heat away from the unit. This is true in general for power electronics. Heat-sink fins that rely upon natural convection of air (the 'chimney effect') function properly only when mounted with the fins vertical. Wall mounting for either concrete or sheet-rock walls uses 4 anchor bolts at the corners of the enclosure. The upper two holes have slip-on capture of bolt heads which hold the unit while the lower two screws are screwed in to their anchors. Be sure to mount the unit on a wall that is open above the unit to allow heat from it to rise. Do not, for instance, mount the unit directly beneath a shelf. (Don't block the invisible chimney.)

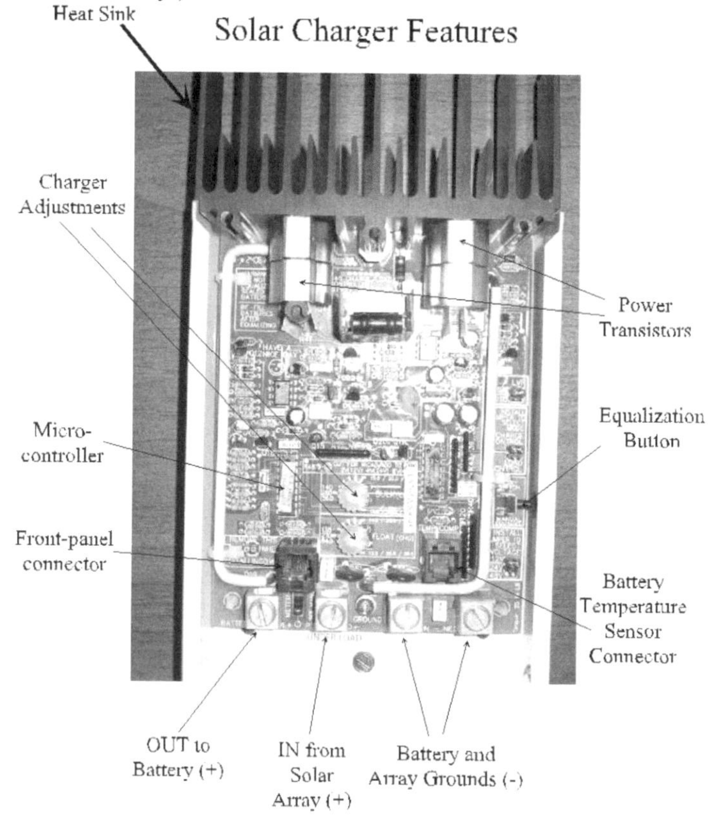

The heat from the charger is mainly dissipated by the power transistors, which are mounted as shown, under the clips holding them to heat sink flanges reaching downward and over the circuit-board. The large house-gage wires running vertically along the sides of the board deliver the high currents to and from array and battery terminals on the bottom of the board to the power transistors at the top. (Ideally, the board design would place these components in close proximity to minimize high-current board wiring.)

The two modular connectors of the kind used with telephones connect multiple wires to the front-panel (left connector) and battery temperature sensor (right). The equalization button allows manual start of a battery-charger equalization cycle, explained under 'Equalizaton'. The two adjustments set charger parameters. Additional jumpers, spaced vertically along the right side of the board, configure the unit to be used for wind generators (LVR: load voltage regulation) or solar arrays, battery-bank voltage (12 V, 24 V, or 48 V) selection, and whether equalization (or LVR) is manual or automatic.

The microcontroller is the command center of the product and is an integrated circuit (IC), one of the insect-looking components. It inputs sensor data, outputs control waveforms to power transistors, and communicates with the μC on the front-panel board. It is internally programmed at the factory and is a custom part. Most or all of the other parts are non-custom and can be bought from electronics component distributors for those who maintain their own electronics equipment.

A connected C-60 charger is shown below. The red (+) and black (−) # 6 AWG wires come into the enclosure through a metal knock-out hole. These large wires are hard to route and extreme care must be taken if the other ends are already connected to the batteries or solar array. Circuit breakers are used for safety to open these connections in a well-configured system. The front-panel is lying with its back side toward the camera, and the connecting cable can be seen. It works installed with either end plugged into either connector. The battery sensor cable also comes up through the knock-out hole and into its connector on the right. To the immediate left of the front-panel connector on the board is another connector with two female sockets. When the plain (non-instrumented) front-panel option is used, a LED is mounted in this connector and is the sole indicator. It displays a series of blinks, from 1 to 5 or stays on all the time. The six states correspond to battery voltages and give a rough indication of the state of charge.

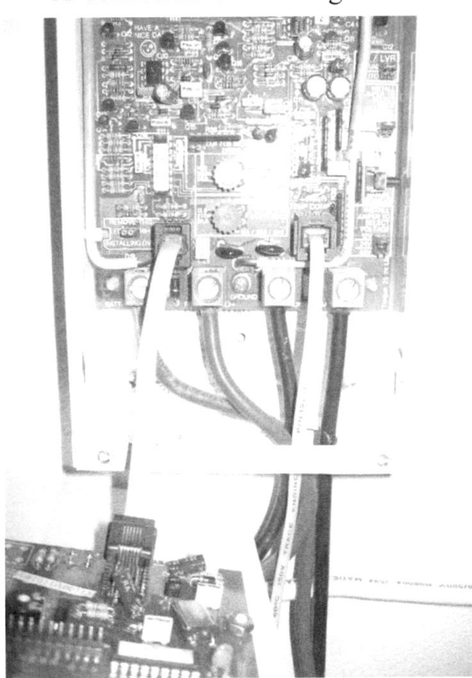

Battery Charging

Batteries are charge storage devices. Lead-acid batteries of either the wet cell (vented) or sealed kinds consist of cells connected in series to provide the specified battery voltage. Charge is replenished by a charger by inputting current (which is charge rate) into these battery cells from the battery terminals. The process is not ideal and more charge must be input than can later be output from the battery.

The simplest charger is a direct connection of solar panels to batteries. This is simple but inefficient and can reduce battery life. If done, then the battery and panel voltages must match. A 12 V battery should be connected to a 12 V solar panel. Then the charging will occur near the max-P point of the panel. However, no charge control exists and a large, well-illuminated panel can overcharge a battery. When this occurs, the battery will lose electrolyte through venting and must be replenished. Do not expect to achieve the specified life for the battery. In emergency situations, this is an acceptable way to charge a battery but it is not a good long-term solution.

The better charger is the *three-state charger*. The state that the charger is in depends on the battery voltage. When the voltage is below the *bulk-voltage threshold*, V_{BULK}, the charger outputs a constant current. In off-line chargers a constant current is essentially achieved, but in solar chargers powered by the vicissitudes of the sun, the value of current can be less than the bulk-charging value of I_{BULK}. The charger outputs a constant current during the bulk-charging state to keep from charging the battery too quickly, which can damage the battery and reduce charging efficiency. For a 12 V battery, typical V_{BULK} (set by one of the two charger adjustments) is about 14.1 V.

Once the battery voltage has come up to V_{BULK}, the charger changes to the *absorption state* where it applies a constant voltage of V_{BULK} to the battery. When the absorption state has been reached, the battery is charged enough so that it will not conduct excessive current if a voltage source is applied to it instead. An ideal voltage source will maintain the voltage across its terminals regardless of the amount of current. Ideally, a dead short across an ideal voltage source will continue to maintain the source voltage across the short, at infinite current. In practice, actual voltage sources can provide enough current to be excessive if not limited.

By maintaining a constant voltage across the battery, additional charge is forced to flow into it and is 'absorbed'. This is the harder charging that is necessary for a fully-charged battery. This state lasts the longest, and during it the charging current will steadily decrease until it reaches some sufficiently small value, I_{FLT}, and the battery can be considered fully charged. When the charger detects that the charging current has dropped below I_{FLT}, it changes to the third state.

The third charging state is the *float state*. The battery is driven with a reduced voltage ($V_{FLT} < V_{BULK}$) to maintain its charged state. V_{FLT} is the other charger adjustment. A typical value for $V_{FLT} = 13.4$ V for 12 V batteries. With time, batteries slowly self-discharge and the float state counteracts that tendency. When sufficient charge has been drawn from the battery, its voltage will drop beneath a threshold, V_{CHG}, and the charger will change to the bulk-charge state and proceed through the sequence of states once again.

Equalization

The three-state charging scheme is widely used and maintains reasonable battery life. However, the cells in a battery are not identical, and 'weaker' cells will not be charged as much as other cells in series with them. Battery cells in a discharged state degrade and this reduces battery life. To compensate for cell differences, a technique called *equalization* is used to partially remedy cell charging differences.

First, the battery is charged with the usual three-state charging cycle. Then an equalization cycle is begun. A voltage greater than V_{BULK} is applied to the batteries. Some cells will overcharge but the weaker ones will fully charge. This cycle is similar to the absorption state except with higher voltage. After equalization, the fluid levels in the cells for vented or 'wet cell' batteries should be replenished. While charging, the fluid level in cells will fall, and after equalization, will be lowest. For any cells with low levels, distilled water is added to bring the levels high enough to cover the plates.

Equalization is a controlled overcharging that depends on replenishment of fluids, something that cannot be done for a closed battery. Consequently, sealed batteries (gel cell and AGM) should not be run through an equalization cycle.

Inverters

Inverters input dc and output ac. The 'inversion' is in going from a constant (or dc) voltage to a bipolar (+ and −, or ac) waveform. Inverters, like solar chargers, are an electronic block in the power system. They are specified by output power, input voltage, output waveshape and voltage, and various other features.

Inverter Features and Selection

Assessment and selection of inverters is based on their more important features and specifications. Here, some of the more important characteristics are surveyed.

Waveshape: Sine-waves are usually preferred, though many cheaper inverters output a bipolar square-wave, referred to in the commercial literature as a 'modified sine-wave' or 'quasi-sine-wave'. They are neither and are simply square-waves. A square-wave contains a sine-wave of the same frequency, but it also contains multiples, or *harmonics*, of that frequency. These harmonic sine-waves cause additional power dissipation (lower efficiency) and electrical noise, and are generally undesirable.

Voltage: The inverter input battery voltage should match your battery bank for battery-bus systems, though a 12 V unit can often handle 24 V.

Power is specified in units of watts or kilowatts (kW), with peak and continuous ratings. The peak rating cannot be sustained for long, usually not for more than one to ten seconds before the power components begin to overheat, leading to inverter shutdown. The peak rating is important for starting motors or other loads that momentarily draw two to three times their operating current when starting. The continuous rating in cheaper inverters is tenuous. Based on your maximum power from load analysis, an inverter rated at 25 to 50 % more power is recommended.

Undervoltage shutdown: When the battery bank is being severely discharged, the inverter will detect a cut-out voltage and turn off to protect both battery and inverter. A typical undervoltage threshold for a 12 V lead-acid battery bank is 10.5 V. For extended battery life, the bank should not be discharged more than 50 % (corresponding to 12.1 V) and the inverter should provide a warning (usually an audible alert) that battery discharge has exceeded this amount. A typical warning threshold is 11.0 V. A12 V threshold for home energy systems would be better.

Outlets: Multiple ac outlets are convenient in that one can be used as input to the load center (breaker box) while others are used in the energy room for lighting. Larger inverters include a 240 V outlet.

Power Indicator - a display often implemented as an LED bar graph that shows roughly how many watts are being output from the inverter. As a simple power meter, it is quite useful in monitoring the state of the electric system and in diagnosing problems.

Efficiency: Switching inverters have efficiencies in the range of 80 to 90 % or more. One major aspect of efficiency, other than the input-to-output efficiency, is the amount of power used by the inverter itself. Generally, as inverter power ratings increase, so does the power used by the inverter. This constant drain on the battery bank can be minimized by selecting an inverter that is not too grossly oversized. If the total combined loading of your system (when everything plugged in is on simultaneously) is, let us say, 350 W, then choice of a 500 W or even a 750 W inverter gives a reasonable power margin. However, if the 750 W unit draws 50 % more current when idling than the 500 W unit, you might want to settle for the 'lighter running' 500 W unit to save power. When calculating loads, be sure to include the inverter itself as a load on the battery.

Noise: For inverters located remotely in an energy building, fan noise might not be a consideration. However, if the energy facility is closer to your house so that you can hear the noise of inverter fans while in bed at night while trying to sleep, then it becomes a consideration - annoying and often overlooked!

Size & Weight: Some inverter-chargers are 'boat anchors'. One 2500 W unit I once owned weighed well over 77 pounds (35 kg) and was as large as a desktop computer. I now use a 1500 W unit (pictured at the beginning of the chapter) made for the vehicle market that does not have the quality of the Xantrex unit but I can easily hold it in one hand. Size and weight become considerations for installation (especially wall mounting) and for humanly moving these units for repair, especially if you are not a 'big hod'. Energy rooms that are space-limited also benefit from a smaller table-mount unit that can be placed within a foot or two (less than a meter) from the battery bank.

Reliability is one of the most important and least specified characteristics of inverters. You find out about it by word of mouth from satisfied or, more often, disgruntled users. In my view as an electronics engineer, I have yet to find an inverter designed to my satisfaction, though they could well be. Why? Price competition drives companies to cut corners in the designs, resulting in less-than-satisfactory products.

Maintainability: Whether you have electronics repair skills or rely upon others in your locale who do, technical documentation of products is necessary - and unavailable from inverter companies except through their authorized dealers. This can be very frustrating because, by forcing most after-sale customers to rely upon the company maintenance system, absurdly high prices can be (and are) charged for repair. I have had to draw out the circuit diagrams by tracing the wiring on the boards of my inverters so that I can maintain them, though most inverter users could not be expected to do this. Your local, preferred electronics repair shop would have to do it (if able) and it is time-consuming. Before buying an inverter, have a clear understanding of how you will maintain it over its product life. I keep two additional units for backup in case of failure, and for the three I paid far less than for a single comparable higher-quality unit.

Inverter Load Instability

Reactive loads can cause inverters to become unstable. Besides CFLs, other reactive loads can also cause inverter instability. This usually occurs at inverter start-up and can be recognized by flickering CFLs, trying to start but not lighting continuously. A well-designed inverter will not show these problems, but few low-cost inverters are. One solution to this problem is to recycle the inverter power switch. If the inverter does not stabilize, then switch out one of the branches in the load center. This can let the inverter voltage come up to a stable value and settle. Then the breaker can be switched in and the inverter will (hopefully) hold stable and continue to operate correctly.

It is not unusual, as happens in houses on a utility grid, for the starting of a motor to momentarily blink the lights because CFLs and LEDs respond at electronic speeds while incandescents (if you are still using them) are an averaging light-emitter and 'filter out' quick blinks. A transient demand of high current on the distribution lines will decrease the distribution voltage momentarily by causing a large voltage drop on the lines. Some inverters are also not capable of following quick changes in output current in their control circuitry and will output a 'dip' in voltage on the distribution lines before they can readjust to the higher current. Only cheap computers without local battery backup and programmed loads (such as bread-making machines) suffer adversely from these quick transients. If they are a problem, the simplest solution is to relocate big motor loads nearer the source to minimize power-line voltage drop. Another solution is to get a box that induction motors can be plugged into that 'soft-starts' the motors at limited current from the outlet.

Inverter Fan Refurbishment

Axial cooling fans that are mounted usually in the rear of inverters are required to provide circuit cooling. These box fans are generally reliable, but they run day in and day out and are mechanical devices. Their bearings eventually fail, especially when supplied with ambient dust, mold, and humidity (or 'dust, must, and rust', the three

omnipresent characteristics of warmer climates). As they begin to fail, they will sound like they are straining, or squeaking.
They can usually be refurbished by removing their mounting screws, noting the mounting configuration for them and the grill. Their two wires are usually color-coded (red for +; black for −) and if they are not long enough to allow you to work on them while connected, unsolder them from the circuit-board. *Be sure to note which holes the red and black wires each go into!* If soldering is beyond your skill range, cut the wires at an accessible point, then later, strip insulation off each end, twist them tightly together, and apply some tape to cover the metal wiring to keep it from shorting to other metal in the inverter. See 'Fan Repair' in *Eco-Electrical Home Power Electronics* for details

.

Inverter Repair Tips

A disassembled 225 W-rated Vector Mfg. Co. VEC034D inverter is shown below.

To repair a failed unit, first check the fuse(s). Automotive blade fuses are used and are mounted on the board near the battery input.

The second most likely failed components are MOSFETs. Check them first visually for any obvious cracks or other signs of failure. Then with a digital multimeter (DMM) set to ohmmeter (2 kΩ range), measure from the front-view left pin to the other two terminals of the MOSFET. This can often be done with the part in the circuit. All readings should be 'open circuit' (blinking reading). Any shorted pair of terminals gives a low reading and indicates a failed part.

If the converter MOSFETs test good, next check the inverter H-bridge MOSFETs in the same way. These components are often not adequately protected by overcurrent circuitry (another design shortcut) and can fail. To replace them, remove them from their heat sink (a nut and bolt or clip) and unsolder the three terminals. Order comparable parts (and some extras for next time) from a supplier such as www.mouser.com , www.digikey.com or www.futureelectronics.com.
Beyond MOSFETs, check power diodes and then the small-signal parts.

Power Distribution and Loads

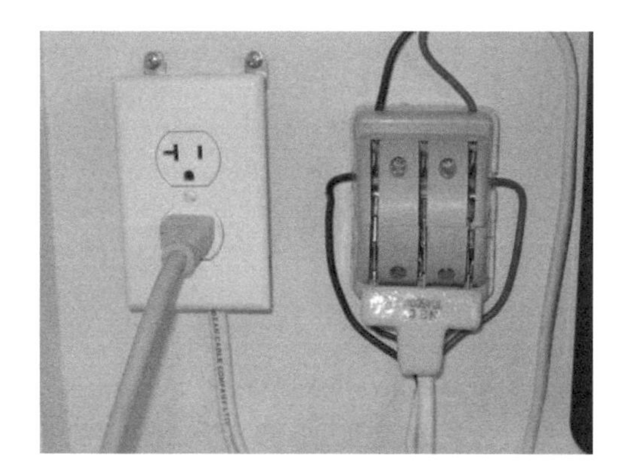

Electric power is usually not used where the power source is located, but is distributed by *wiring* to loads which use the power. The standardized method of connecting loads to the lines is by means of standardized *plugs* and *outlets*. In different regions of the world, the standards are different, as any world traveler knows. A North American two- or three-pronged plug will not plug into a European outlet socket - and just as well, for the consequences would be undesirable! In this book, the North American power distribution system is assumed, with its 120 V, 60 Hz sine-waves, though the principles are the same globally. In home systems, the waveshapes from inverters might instead be square, and the amplitude of the voltage might be less than 170 V. (See the chapter on 'Inverters' for why this matters.)

Average Power

Some people have trouble thinking of power as the energy rate. When given a number such as 100 watts, some will ask: Is that 100 watts per day? No, it is 100 W all the time. The energy delivered at the rate of 100 W over 24 hours is (100 W)·(24 h) = 2,400 W·h, the amount of energy that is delivered at a 100 W rate over a day.

Sometimes advertising information causes confusion. Appliances are rated on yellow tags with their energy use per year in kilowatt·hours/year. This is a rating of average power (not energy), which can be converted to the more useful unit of watts for home systems off the utility grid. There are about 365.25 days times 24 hours/day, or about 8766 hours/year. Then

$$\frac{\text{kilowatt}\cdot\text{hours/year}}{8766\ \text{hours/year}} = \text{kilowatts, average}$$

This can more conveniently be expressed in average watts instead as

$$\frac{\text{kilowatt} \cdot \text{hours/year}}{8.766\ \text{kilohours/year}} = \text{watts, average}$$

A Maytag refrigerator is rated at 447 kW·h/yr. Dividing 447 by 8.766, it uses an average power of 51 W. The power is next to nothing when the refrigerator compressor is not running but it is perhaps 204 watts when it is. Because it uses only 51 W on average, then it runs only (51 W)/(204 W) = ¼ of the time. Over time, the on and off powers average out to a 51 W usage.

The Power Budget

The distribution and load subsystem can be described as a list of loads in a power-budget table such as the following one.

Load	Operating Power, W	Average Power, W
Refrigerator with freezing compartment	200	50
CFL Lights	50	6.5
Water pump	350	3.5
Cooling fans	50	15
Equipment (inverter, consumer electronics, computers)	350	25
Total power	1000 W	100 W = 2.4 kW·h/day

A North American household typically uses an average of 750 W to 1 kW. However, 200 W can be achievable with modest accommodations in lifestyle. In developing countries, 400 W might be expected by urban residents. Of equal significance is industrial demand for those who work at home or have a shop or laboratory.

Peak Power

If all loads of a system were to be turned on at the same time, the maximum or peak power demand would occur. By including the center column in the above table for the operating power of loads - the power they use when running - then the sum is the maximum or peak power that the system can be expected to have to supply. It is unlikely that all possible loads would be on simultaneously. Consequently, some judgement is required in assessing what the actual maximum might be.

Electrical Outlets

Electrical outlets are standardized in North and Central America for 120 V ac. While the standard for socket pin functions should be followed, for home power distribution you can decide on color coding where there is no clear standard. The three-conductor wall socket convention I use is as follows:

Left slot (long): neutral; black wire
Right slot (short): hot; red or white wire
Lower, round slot: safety ground; green

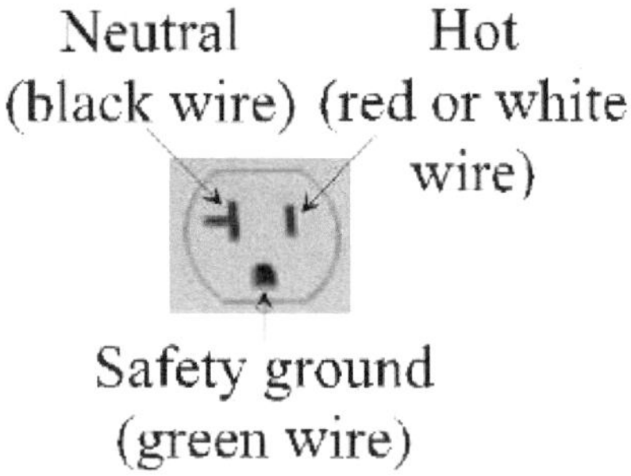

Neutral (black wire)
Hot (red or white wire)
Safety ground (green wire)

In North America, black is usually used for the hot terminal, though in electronics, it is more indicative of a ground node. The neutral line is connected at the source to the source terminal that is connected to earth ground. However, both neutral and hot lines should be considered dangerous. The safety ground connection is added so that the metal cases of equipment can be connected to it. Then, if the hot line were to short to the case, it would short the power-line and open a breaker. Without this safety feature, a human could instead become a conductive path from the faulty equipment enclosure (shorted to the hot line) and ground, causing possible electrocution.

Lighting

Lights are generally used in two ways: spot and illumination. Spot lighting, as done by flashlights, under-counter, and track lights, constrains the light to a relatively small 'solid' or 2-D cone angle, concentrating it on where it is needed. Illumination is used to light larger volumes such as whole rooms.

Optimal placement of lights in a house is often a non-trivial problem. Pantries with stacked shelves ideally have a light under each shelf, but this can be excessive in off-grid homes. A few well-placed LEDs can often light dark corners where area lighting is inadequate. Before installing LEDs, test lighting requirements with a small LED flashlight placed where LEDs might be installed. LEDs have the advantage of being small and capable of being mounted in cramped places. CFLs are better used for area illumination. Use of floor or countertop lamps allows them to be moved to where light is needed, though they tend to be cumbersome to move for daily use. By beginning with mobile lighting, patterns of use can be identified and then addressed with stationary lamps.

Motors as Loads

Motors are commonly found in 200 W to 1 kW sizes in homesteads. They are used in water pumps, as compressor motors in refrigerators and air conditioners, hand tools, and in powering large fans or blowers. A small variant of them is the shaded-pole motor found in desk or vent fans. When motors start, they draw a transient current appreciably more than their running current. Starting current can be over twice the running current. As motors reach operating speed, their current demand drops and becomes steady. The starting current must be provided by the inverter or other source and should be taken into account in the power budget. This is the main reason for considering inverter peak power ratings. Inverters must be oversized from the continuous load ratings to handle transient currents.

Other kinds of motors such as brush motors are found in drills or other hand-held power tools, kitchen appliances, and consumer electronics such as VCRs. A newer variant is permanent-magnet synchronous (PMS) or 'brushless dc' motors. These motors last longer and are more reliable than brush motors because they have no brushes to wear out and are electronically driven. They are somewhat more expensive and have yet to become common in household electric appliances.

The use of field-oriented motor-drives, especially for induction motors, controls starting current and drives the motor efficiently, thereby reducing overall power requirements. These drives are now common in industry but are still too costly for consumer markets. However, the homesteader can acquire small surplus vector motor drives for home use.

Electronic Loads

The power-line connects to the input of the power supply in electronics equipment. There are two kinds of supplies. The first and older kind uses a relatively large power transformer that operates at the power-line frequency. It transforms the line voltage down to a lower voltage compatible with electronics circuits. This kind of supply will dissipate more power, lost as heat, with a bipolar square-wave waveform than a sine-wave.

The second and newer kind of power supply is the *switching converter*. These power supplies do the voltage conversion using electronics that generates a frequency much higher than the power-line frequency so that a much smaller transformer can be used to down-convert the voltage. They are also more efficient. Some switching converters have an input stage called a *power-factor controller* (PFC) that makes the converter input appear purely resistive to the power-line. This helps to relieve the problem caused by reactive loads by drawing power from the power-line waveform over its complete cycle rather than only at its peaks, as do most transformer-input power supplies.

Electronic loads are usually more fragile than motors because they are subject to failure due to overvoltage. This is ordinarily not a problem for residential utility-grid loads, and is more likely to occur in home electric systems due to inadequate inverter voltage regulation or the variation in speed of engine-driven generators. Overvoltage occurs transiently as loads switch in and out, causing current demand to change quickly. Occasional overvoltage transients are controlled by the use of transient voltage suppressors (TVSs). They are commonly found in power strips and wall-mount power cubes. They have a life of only a few overvoltage events, however. TVSs can also be purchased and wired across existing wall sockets, between the hot and neutral connections.

Reactive Load Instability

Compact fluorescent lamps (CFLs) have their own self-contained inverter electronics that drives the lamp itself. The input terminals of a CFL - the two connections of the screw-in bulb base - present a resonant circuit to the inverter. The CFL electronics drives the lamp using a resonant circuit and the current of this resonance is drawn from the power-line. The CFL and power-system inverters operate at different frequencies and this can cause adverse interactions and instability.

Some inverters become unstable in their regulation of ac output voltage for reactive loads. The effect is that CFL lamp loads will blink on and off at a rate of two to five times per second, as the inverter tries to re-start and stabilize. The Vector Mfg. Co. VEC050D is an example. Sometimes it will not output power at power-on when loaded. If this occurs, switch off one or more load breakers until one of the distribution branches shows power. Then switch in the other breakers, one at a time. The fan speed is an approximate audible voltmeter. When it returns to normal speed, then switch in the destabilizing distribution branch.

Some inverters regulate poorly or shut down when the input voltage is high, such as over 13.8 V. Poorly-designed overvoltage protection circuits do not shut down cleanly and will interact with inverter regulation, causing similar symptoms to those of reactive loads.

Inverters that output bipolar square-waves will often need to drive equipment with PFC converters. This should not be a problem in that the waveshape presented to the PFC should not affect its function. PFCs are usually beneficial in minimizing load-caused inverter instability.

Some loads, such as uninterruptable power supplies (UPSs), used to insure continuous power to computers, depend upon sine-wave power inputs to work properly and often will not function well, if at all, on square-wave power.

High-Resistance Plugs

A dramatic cause of failures in power distribution is high-resistance connections. A power plug that has poorly-connected wires in the screw terminals can develop a high resistance. By Watt's Law, the power dissipated in a resistance is $i^2 \cdot R$, where R is the resistance and i the current. The excessive dissipation will cause excessive voltage drop and low voltage to loads. Constant-power loads will increase current to compensate for the lower voltage and a circuit breaker will turn off. When breakers 'trip' (turn off) without any detectable reason, do not overlook connectors that are charred, partly melted, or feel hot to the touch.

Gasoline and Diesel Generators

Sine-waves are generated by electric machines called *generators*. All motors are generators and all generators are motors. This is why they are sometimes referred to by the more general name of *electric machines*. If the input power is electrical and the output power is mechanical, the electric machine is operating as a *motor*. If the reverse is true, it is a generator.

The generator shaft is connected to that of an internal-combustion engine, as shown above (with generator to the right). A smaller engine-generator coupling is shown below.

The engine is usually a gasoline or, preferably, a diesel engine. Together, the pair is a 'generator set', 'genset', or simply a 'generator', with the understanding that this means both.

If generator size, in maximum output power, is chosen to accommodate large anticipated loads such as welders or clothes dryers, and these loads are run only occasionally, then a large generator will run most of the time under light loading - that is, at low output power. Consequently, the efficiency at minimum output power, or *zero-scale*, should be as high as possible. Diesel engines are typically more efficient than gasoline engines, especially under light load.

Diesel engines also do not have the problem of spark-plug fouling when the carburetor is misadjusted for too rich of a burn (too much gas per air), though diesel fuel injectors can occasionally require servicing. Diesel fuel is comparable in price to gasoline. Diesel engines are sturdier and weigh more, making them both less susceptible to damage and harder to move. They also cost somewhat more.

Major components of a gasoline generator are shown below.

In the upper-left picture, the gas cap has been removed from the tank for filling. It is sometimes easier to fill the tank when the filter cup is tilted at an angle, as shown, to allow air to be pushed out of the tank by the incoming gas. Gas gages of the kind shown are marginally accurate and sometimes stick. Tap the gage with your fist to verify that the indicator has moved to its correct position.

The upper-right picture shows the engine switch. Turn it on before pulling the starter cord (mid-left picture). Pull the cord gently out at first to rotate the ratchet mechanism to where it pulls harder. This is the position where all ratcheting slack is taken up and the crankshaft will be turned for the full range of the cord pull. The engine should start (most of the time) in less than a full pull. If not, check the choke setting (the lever above the air cleaner) and that the fuel valve is open.

The middle-right photo shows the air cleaner to the left of the pull-cord. Fasteners at top and bottom of the rectangular box snap off when pulled and also snap back on.

The lower-left photo shows, from front to rear, the speed controller, or *governor*, spark plug cap, and the top of the muffler. The speed controller has a lever and an adjustment screw for setting the engine speed. This can be set either according to the generator output frequency (60 Hz) or, more easily in most cases, by the output voltage.

Gasoline and Diesel Generators

Some units have a voltmeter on the panel, as shown in the lower-right picture. It is best not to adjust the speed setting too frequently. The voltage will vary depending on load and also as the engine warms as it runs. The voltage should be in the range of 110 V to 125 V rms. If it is too high, equipment might be damaged when line transients occur. If it is too low, constant-power loads (such as newer equipment with switching power converters) will draw more current and the system could be over-current and activate the generator circuit breaker. Two of them are shown on the control panel (lower-right photo), one for each outlet.

The two generator outlets (unlike inverters) are usually *not* wired in parallel but are opposite in phase. If they were connected in parallel from the front-panel, the generator would be shorted and damage would occur without the breakers. To achieve full-scale output power, both outlets must be used. One way this can be done is to use the two outputs as sources for two load distribution branches. This scheme works optimally when the anticipated loading of both branches is about equal.

A better solution for 120 V-only operation is to reverse the phase of one of the 120 V generator windings at the connector. This also requires modified front-panel wiring. For both 120 V and 240 V options, mount a 120 V/240 V switch and connect windings to it from generator and plugs, paying careful attention to phase.

Generator Maintenance

Engine or electric machine repair is beyond the scope of this book. Regular maintenance and some tips on engine behavior, can be useful, as given in this checklist of regular maintenance items:

- Oil should be changed regularly. Newer generators have low-oil sensors and if the oil level is too low, the engine will not start. Also, do *not* over-fill the crankcase with oil either. This can also lead to engine problems. If the oil appears dirty, it is probably time to change it. An easy way to do this is to suspend the generator on wood beams, as shown for the 5.6 kW gasoline generator in the opening photo of the chapter. Then an oil pan can be inserted underneath the engine to catch the oil when the plug is screwed out. In tropical climates, use of 40-weight oil is recommended.
- Spark plug (for gasoline engines): If the engine runs rough - that is, it sounds like it is missing power cycles - pull the spark plug out with a wrench that is supplied with the generator. A socket wrench will usually work. Clean metal surfaces with a file or grit paper and adjust gap. Typically it should be somewhat less than 1 mm. Clean carbon off the porcelain insulation too. If the metal electrodes appear to be substantially eroded, replace the plug. Clean the metal ring the plug contacts. Seat the plug cap solidly back onto the plug after it is screwed back in place.
- Air filter: This sponge of foam material is cleaned by soaking it with gasoline, squeezing it out, letting the gas evaporate from it and reinstalling it.

- Mufflers collect soot from the burned combustion products and must periodically be dismantled. Soot can be tapped out once the muffler has been removed from the exhaust manifold and mounting brackets.
- Fuel filters collect sediment over time and should occasionally be checked and cleaned. This applies to the tank too.

These are periodic maintenance items. Beyond these are tappet and timing adjustments, and more serious maintenance such as valve grinding (for reseating) and piston ring replacement.

Plan to take your generator set to a competent mechanic by putting in place the mechanisms for transport of the generator. For smaller generators, such as the 5.5 hp unit shown above, one or two persons can carry and lift it into the back of a vehicle. For heavier generators (such as the 5.6 kW unit shown in the first photo), plan on having the help of another one or two persons for loading and unloading it from the transport vehicle, or acquire a hand truck.

Wind Generators

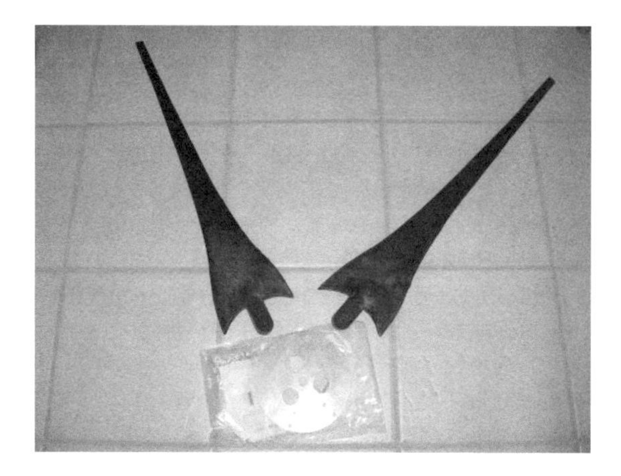

A complementary source of power to sun and petrol is wind. At low latitudes there is not much wind except at sea and on hilltops, but there is usually plenty of sun. In the temperate zones, the wind is a more plentiful energy source. In many places, wind blows in the evening and at night, providing a complement to the sun. A modest wind generator can keep batteries charged and run the few loads during the night such as cell-phone chargers and an occasional light or water pump. The key determination is whether indeed there is any wind where you live. This is most readily determined by paying attention to wind daily during both day and night for a year or more.

Large wind generators for home use, such as the Bergey 7.5 kW unit, require a crane to lift in place. They are too much for anyone without a farm shed of heavy moving equipment. What seems to be optimal for both construction and maintenance are generators with output power in the range of 1 kW to 3 kW, suspended on a ground-hinged, rotating, winch-erected pipe tower with guy wires (adjusted by carbuncles) terminated on concrete pads. In this power range, the assembly can be handled by one or two persons without the need for large equipment. The tower can be raised and lowered using a winch. For larger power generation, multiple smaller towers are preferable for maintainability and power availability.

The generator can be acquired from automotive sources such as a 1 kW truck alternator, a used diesel or gas electric generator, or even a motor. Any motor is also a generator, but depending on type, the electronics required to use it as a battery charger will vary. (Some useful wind generator electronics has yet to be designed and offered commercially.) Induction motors are in this category. Permanent-magnet synchronous (PMS) motors, also called 'brushless dc' motors, are better suited to the application. Neither PMS nor induction motors have any brushes, commutator bars or slip rings, though their interface to a pole fixed with respect to the ground will require that the electric lines from them have enough slack to allow the generator assembly to rotate with the wind.

Besides the tower, generator, electric cable, and junction box at the base of the tower, the other components of the wind generator are aerodynamic and mechanical. The required functions are

- Propeller blades, hub, and gearing to generator shaft, if needed
- Control surface (a tail) that points the propeller into the wind
- High wind-speed protection mechanism for pointing the propeller out of the wind and perpendicular to it
- Bearing for allowing the assembly to rotate with the wind

Blades are the most difficult component to make and are best purchased. Not only must their shape be right for high efficiency, the blades must be closely matched for rotational balance and force distribution. The plastic blades, with hub, shown above are part of a six-blade set that sells commercially for about $100 US (2006). (See www.survivalunlimited.com for Eagle blades.) Tails and other mechanical components are also commercially available.

Blades should be matched to the generator for correct speed range and torque. Larger-diameter blades turn at lower speeds and are a better match for lower wind speeds. Most wind generators begin to rotate and produce power at a wind speed of about 5 to 7 miles per hour, or about 10 meters per second. (1 mile/h = 2.237 m/s) Generator output voltage and frequency vary widely with wind speed and cannot be used directly for power distribution. Conversion electronics is required. Some solar chargers are designed to also be wind-generator load controllers, for diverting excess power from the charger.

Streams and Home-Made Ethanol

Energy Generation from Streams

Large-scale hydroelectric potential has been largely exploited by utility companies, though smaller creeks and rivers have not. Rural and even suburban dwellers not uncommonly have creeks with ponds on or near their property. The feasibility of 'microhydro' generation, as it is called, depends on the power in a stream. This leads to some basic fluidic power calculations.

The basic equation for *flow work* is: $W = P \cdot V$, where P is pressure (not power) and V is the volume (not voltage) of fluid displaced at that pressure. The power equation is therefore

$$\dot{W} = P \cdot \dot{V}$$

where the dots over W and V indicate time rates. $\dot{W}$ is the symbol for power, where W is work or energy, and the dot over it indicates the rate of energy, which is power. Pressure drop from an open stream is due to a change in height due to gravity. The pressure difference (or *head*, as civil engineers call it) is

$$P = \rho \cdot g \cdot h$$

where ρ = fluid density, g = gravitational acceleration (32.3 ft/s^2 or 9.81 m/s^2), and h is the drop in distance. The units of each quantity need only be consistent so that the final result for pressure is in the desired units. Stream power is then

$$\dot{W} = \rho \cdot g \cdot h \cdot \dot{V}$$

Water has a convenient metric density of

$$\rho(\mathrm{H_2O}) = 1\,\mathrm{kg/l}$$

where a liter, l, is 1000 cm^3 = 10^{-3} m^3. (A US gallon is 3.785 l.) Then with earth surface gravity, stream power is

$$\dot{W} = (1\,\mathrm{kg/l}) \cdot (9.81\,\mathrm{m/s^2}) \cdot h \cdot \dot{V} = 9.81 \frac{\mathrm{W}}{(\mathrm{l/s}) \cdot \mathrm{m}} \cdot h \cdot \dot{V}$$

For a 1 m drop of a water stream flowing at 102 l/s, the stream power is 1 kW. To relate this in English units of gallons/minute (gpm),

$$1\frac{\text{gal}}{\text{min}}\cdot\frac{1\,\text{min}}{60\,\text{s}}\cdot\frac{3.785\,\text{l}}{\text{gal}}=0.06308\,\text{l/s}$$

Or

$$1\ \text{l/s} = 15.85\ \text{gpm}$$

Consequently, it would take 1617 gpm over a 1 meter drop (about 1 yard) to produce 1 kW, a considerable flow. A high-pressure, low-flow example of 1 kW - applicable in hilly terrain - is 162 gpm falling 10 m, or about 33 ft.

To gain further insight into the flow rates required, 100 l/s is 0.1 m^3/s. This is a stream 0.5 m deep by 2 m wide flowing at 0.1 m/s - a common size for a creek. A 1 m fall of the creek provides 1 kW of input power to a microhydro generator. Optimized matching (not unlike gearing or electronic impedance matching) for flow volume and speed is required of the impeller to maximize efficiency. Beyond impeller matching, the electric generator will have a typical electric-machine efficiency in the mid-90 % range.

One additional consideration for stream power is storage. How much energy can be stored in a pond? A rectangular pond of 100 m by 50 m by 2 m depth contains 10^4 m^3 or 10 megaliters (10 Ml). At 100 l/s, the time it would take to drain it is:

$$t=\frac{10^7\,\text{l}}{100\,\text{l/s}}=10^5\ \text{s}=27.8\,\text{h}$$

A pond the length of an American football field, half the width, and the depth of a tall person is not very small, yet for a 1 m waterfall out of it, it would provide 1 kW for little more than a day. With these numbers, there is no point outfitting a drainage ditch to provide power. Yet a year-round stream of common size is worth considering.

Energy Generation from Home-Made Ethanol

For those in warmer climates, or in sugar- or corn-producing areas of the world, we now proceed somewhat outside the usual range of possibilities to consider family-sized ethanol production. The family of Dr. Eugene Schroder of Colorado, the veterinarian and leader of the tractorcade political protest that tied up government Washington streets with farm equipment, has produced commercial amounts of ethanol in the processing plant that they built in their barn. Homesteaders are usually not inclined to become chemical engineers for this option, though those willing to build a home electric system usually do not artificially limit their skill-sets.

Streams and Home-Made Ethanol

While building such a plant is a nontrivial exercise, the Schroders are not chemical engineers either. One of them, Micki Nellis, has written a very detailed book/CD on just how to do it, titled *How to Make Alcohol Fuel.* You need not necessarily grow your own crops from which to derive ethanol, for unless you are also a farmer, crops are better left for economy-of-scale specialization. If you are urban, you can still get there by acquiring and recycling cooking oil used in restaurants. The organic input material goes through three processing steps, as shown below.

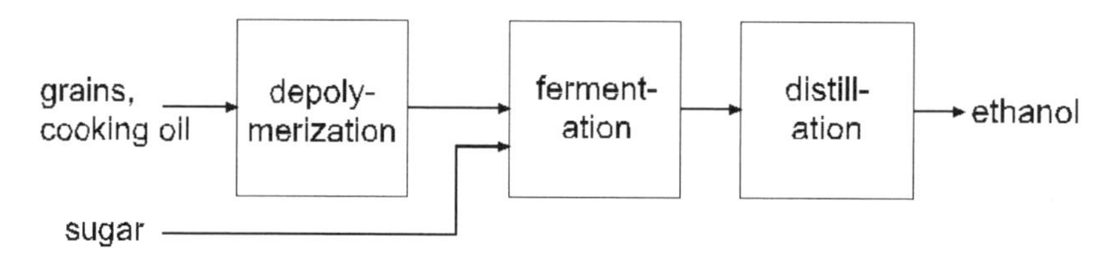

An enzyme is used to break down grains and oils for the fermentation process, which results in ethanol in water. The water is removed through distillation. If you are blessed to live in an area that produces sugar cane, the first process can be skipped. Sugar is immediately fermentable. This process is like that used by breweries, and with microbreweries sprouting up in many places, another approach is to collaborate with a local brewer to make ethanol on weekends or between batches of beer.

A ton of sugar can be produced from 10 tons of cane. (The residue, called *bagasse*, can be burned to run the processing plant.) Cane growing wild will produce 30 tons per acre. With fertilization, 50 tons per acre is achievable. To be conservative, assume 30 tons per acre. Then one acre yields 3 tons of sugar. The sugar-to-ethanol yield is 50 %. (This agrees with Nellis's ethanol/sugar mass ratio of 0.511.) Then one acre yields 1.5 tons of ethanol. The density of ethanol is 6.6 lb/gal = 0.793 kg/l. Then one acre of cane will produce 1720 l. The heating value of (or amount of energy in) ethanol is 23.53 MJ/l (89.0 MJ/gal), resulting in an ethanol energy production per acre of cane of 40.48 GJ = 11.25 MW·h.

This is enough energy to supply 1 kW average power for 1.283 years - about right for a family household with some additional energy use and accounting for conversion inefficiencies. Conclusion: one acre of sugar cane can supply one middle-class family for a year with energy. A direct alcohol fuel cell, or a diesel-, otto-, or stirling-cycle engine must be used to convert it to mechanical power, then a generator to electricity. The heat engines are available technology and the direct-alcohol fuel cell is impending.

What is the cost of ethanol production from sugar cane? At 2006 Belizean labor rates, a cane cutter would cut an acre for $75 US. Brazil has refined the technology, from sugar factories to gasahol plants, and has offered the technology to Belize, a country not far from the US Caribbean coast. An abandoned sugar production facility awaiting capitalization could be an opportunity. Perhaps a few entrepreneurs should get together and solve the energy problem for their network of friends and associates by capitalizing such a plant near sugar-cane fields and providing their people an independent supply of ethanol. The sugar cane industry could certainly use the diversification and the need for energy exists.

Acknowledgement: thanks to Peter Singfield for the rough calculations in his email of July 4, 2001, on ethanol production from sugar cane fields in Corozal.

Power System Design

Now that the components of a home electric power system have been presented, we can put them together to form a system. The 'conventional' power-system design was given as a block diagram on page 2. You do not need to be an engineer to design this system at the component level. Most of what is involved is calculation of the *size* - the ratings - required of various components. That is what will occupy us here. However, before launching into that exercise, it is noteworthy that an improved power-system design is the *high-voltage dc* (HVDC) system, shown as a block diagram below.

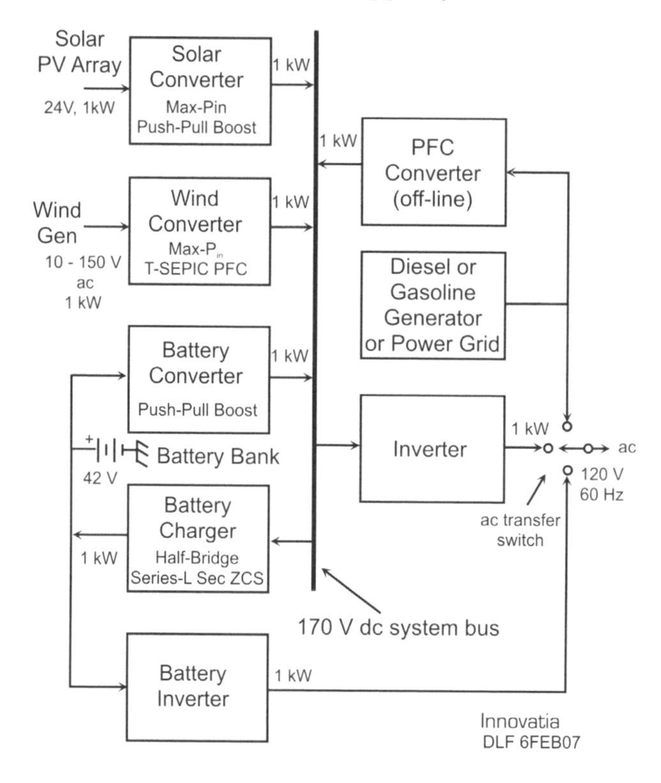

In this 170 V bus system, some of the components of the conventional system are taken apart and made separate components. For instance, the solar charger of the conventional system is decomposed into a solar converter and a battery charger. The inverter is decomposed into a battery converter and the inverter itself. The components are now oriented around a 170 V bus instead of the battery bank.

This higher voltage not only reduces wiring size, but it also increases system efficiency. In the conventional system, all output power - power to the loads - had to come from the battery bank through the charger. In the 170 V dc bus scheme, only the power needed to charge the battery bank goes through the charger. Additional load power is taken directly from the converter sources as they output power to the dc bus. Any output power during generation from one or more of the exterior sources bypasses the battery bank and avoids the losses in efficiency in going through charger and battery converter.

Estimating Load Demand

The first step in power-system design is to begin with the loading and estimate how much power will be required. Work out a power budget. (See the 'Power Distribution and Loads' chapter for details.) Both average and peak power must be considered because the system must not only be able to supply the average over the long term, but also peak power over the short-term. The inverter and also the battery-bank wiring must be able to supply anticipated peak power. For identifiable tasks such as welding or operating a large machine, the system can be switched to a different source such as a generator capable of supplying more power than an inverter. However, for intermittent loads continually plugged into the power-line, if all of them run at the same time, the inverter must be able to supply sufficient power. For instance, if both water pump and refrigerator run times overlap, the inverter must be rated to handle the sum of their powers.

To calculate load demand, make a budget table (see 'The Power Budget' on page 57) listing all loads that the system must be able to supply, their power ratings when operating, and a third column for the fraction of time they will operate. A fourth column, average power, is the product of the operating (or peak) power and on-time fraction. The totals from this table will be used to size various system components.

Distribution System Sizing

House wiring is usually also 'given' based on house design. If wiring is incomplete or the system is being designed along with the house, then the wiring scheme for distribution will be part of the design. The lengths and anticipated loads drawing current from the various branches of the distribution system need to be determined. Then from the wire table calculations in the 'Basic Electricity' chapter the required wire sizes are found. These kinds of calculations also apply to the solar PV array wiring.

Solar Array Sizing

The solar array is rated in peak power as the sum of the power ratings of the individual panels in the array. Depending on one's location, the average fraction of the day that has sunlight varies from nearly 12 hours at the equator to 5 hours in winter in the temperate zone (45° latitude). In Belize (at 17°), 10 hours is an approximate value.

Then the fraction of day that has sunpower is $10/24 \approx 0.41$ or about 40 % of the day. When the low angles of the sun at dawn and dusk are taken into account, even for trackers, the effective fraction of sunlit day is about 1/3. Then consider cloudy skies, and solar charger and inverter inefficiencies, and this fraction reduces to 1/4. For a load requirement of 100 W, 400 W of panel are needed.

An array that is fixed in position (no tracking) must be 1.57 ($\pi/2$) times the size of a tracking array for comparable daily energy output because of the non-perpendicular angle (except at noon) of the sunlight impinging upon the panels. If the panels are fixed and not on trackers, then this value increases by 57 % to about 628 W. Commercially available panel sizes would probably round this down to 600 W.

A general rule for rough calculations is that the panel power should be five times the average load power. This is not highly accurate though it provides an approximate number when one is needed quickly.

If the 600 W panel is rated for 12 V output, then the peak charging current available from it is, by Watt's Law, $600\text{ W}/12\text{ V} = 50\text{ A}$. On average, it provides $100\text{ W}/12\text{ V} = 8.33\text{ A}$. Over the day, the average charging current is $(50\text{ A})\cdot(10/24) = 20.8\text{ A}$ These numbers are approximate because they assume a constant 12 V output (which varies with illumination). However, the voltage and current are relatively constant almost all of the time, and the average power will be their product.

Battery-Bank Sizing

The one big design decision for sizing a bank of lead-acid batteries is the discharge depth. A typical choice is 50 %. This value is a tradeoff between number of life cycles and total energy storage during battery life. To achieve this, whenever battery voltage decreases to 12.1 V and no solar or wind energy is available, generator charging is required. One rule of thumb is to check the battery voltage in the evening a couple hours before shutting down for the evening. If it is 12.2 V or lower, run the generator and off-line charger(s) for an hour or two to boost the charge for overnight, until the morning sun can resume charging.

The battery-bank charge capacity is its size. A system using an average power of 100 W will, in one day, require an energy of

$$(100\text{ W})\cdot(24\text{ h}) = 2400\text{ W}\cdot\text{h}$$

It requires $2400\text{ W}\cdot\text{h}/12\text{ V} = 200\text{ A}\cdot\text{h}$ of charge each day. At 50 % discharge, the battery bank size is effectively half its rated charge capacity. Then its rating must be twice the daily charge requirement, or

$$200\text{ A}\cdot\text{h}/0.50 = 400\text{ A}\cdot\text{h}$$

Using 200 A·h batteries, then 400 A·h/200 A·h = 2 batteries. As batteries age, their charge capacity decreases and this battery-bank sizing should be considered an absolute minimum.

Solar Charger Sizing

The charging current requirement of the solar charger assumes charging over the whole day because solar-panel sizing was based on a full day of sunlight. A three-state (or three-stage) charger will charge at a constant current until the battery voltage increases to a given voltage, then charges at a constant voltage. During the constant-voltage or absorption state, current decreases until it reaches a low value. This can take a long while. Then the battery is charged and the charger keeps it 'topped off' at a trickle-charge current.

Solar chargers are often designed for this kind of charging but the actual charging dynamics are usually determined by the sunshine itself. As clouds come and go, charging current rises and falls with illumination. It is only after nearing full charge that the absorption and float control is significant. In the 'Battery Banks' chapter, battery charge rates are given. For wet cells, the maximum charging rate is the battery charge rating divided by 4 hours. For the previous example, 200 A·h/4 h = 50 A. This value is well above the 20.8 A of charging current that the panels are capable of sourcing.

For a daily requirement of 200 A·h of charge and 10 hours of sunlight, then the charging rate, if it were constant current, would be:

$$\frac{200\ \text{A}\cdot\text{h}}{10\ \text{h}} = 20\ \text{A}$$

This is consistent with panel sizing and its maximum current of 20.8 A.

What might have complicated these calculations is the fact that while the sun is charging the batteries, they are being discharged by the inverter driving an average 100 W of loads. However, the previous calculations took this into account. The total energy required by the batteries includes the energy for the entire day. The average power applies all the time, 24 hours per day. This includes the 10 hours of sunlit day. Charging and energy requirements were based on the average power and thus all power use was included.

Inverter Sizing

The inverter input voltage must be compatible with the battery bank. The upper limit on input voltage is the voltage rating of the power transistors (usually MOSFETs). These devices are usually not less than 40 V in their breakdown rating, but some safety margin is required to account for voltage transients, mainly from chargers. Also, the typical push-pull converter circuits used in many inverters double the MOSFET voltage requirements. If 25 % margin is provided, then the maximum input voltage for 40 V MOSFETs is (40 V/2)·(0.75) = 15 V. A 24 V input will overvoltage most inverters that are rated by the manufacturer at 12 V. Overvoltage shutdown provides protection up to the MOSFET voltage ratings.

One other complication is the control circuitry. If it is not designed to regulate at the higher input voltage, then this can cause potential inverter failure or, if the loads are overvoltaged due to lack of control, load failures can occur. CFL loads are particularly vulnerable to overvoltage failures.

The other major inverter sizing issue is the output power. From the load budget, the inverter power rating should exceed the average load power to leave a safety margin for the inverter. How much depends on the quality of the inverter design. Commercial high-volume inverters found in department or automotive supply stores are designed for low price and corners are cut in the design. I have reverse engineered over a half dozen 'consumer-grade' inverters and all of them have design shortcuts that can cause failures. For these inverters, a 50 % margin is not excessive. If a load budget has a maximum average power of 750 W, a 750 W-rated inverter will likely fail within a year. Even a 1500 W unit is not 'bulletproof' but would be recommended for a 750 W budget.

For inverters designed specifically for alternative-power systems, such as those made by Xantrex or Outback, a 10 % margin should be sufficient. Some additional margin accounts for unusual operating conditions that might exceed the manufacturer's conditions for the ratings. It only takes one failure event to leave you without power from that unit until it is repaired. If it is your only inverter, you will be able to empathize with the lifestyle of the Amish for a while. Keep a back-up, even if it is only a low-cost consumer-style inverter.

If peak power lasts for only a few seconds, then most inverters can handle the transient overload. The inverter rating should also be within the peak power. If it is more than twice the average load-budget power, it would be best to scale up to a higher-power inverter.

The disadvantage of larger inverters is that they dissipate more power. An inverter itself uses some power and this power should be included in the load budget, though it is a small fraction of total power use. A choice of inverter power rating should also keep this in mind. A bigger inverter is not necessarily better.

Another factor in choosing an inverter is the waveshape. Sine-waves are preferred, though most loads, including the newer electronic switching converters - those 'wall warts' or table-top power-supply boxes that are found nowadays among printers and laptop computers - might operate even somewhat more efficiently on bipolar square-waves than sine-waves. For older devices with line-operated power transformers, additional power loss will occur. Some very marginally-designed consumer items might even fail. And radio-TV noise in audio is greater from bipolar square-waves.

Power Facility Design

Electric power system components must be placed somewhere. Of necessity, the solar array must be outdoors in an unshaded area. Also of necessity, the battery bank, inverter, and solar charger must be in close proximity. An electric power room, closet, or 'facility' of some kind is also needed to house the system. One such facility is shown below.

This particular facility is a room jutting out from the house with a stairwell behind it for placing the gasoline generator and accessories. The room should be well-ventilated because of the batteries and also to keep power equipment cool. It should also be clean and not apt to become dusty. A concrete patio or chipped stone (both used above) will aid in dust control. With minimal external dust, windows can be left open for ventilation.

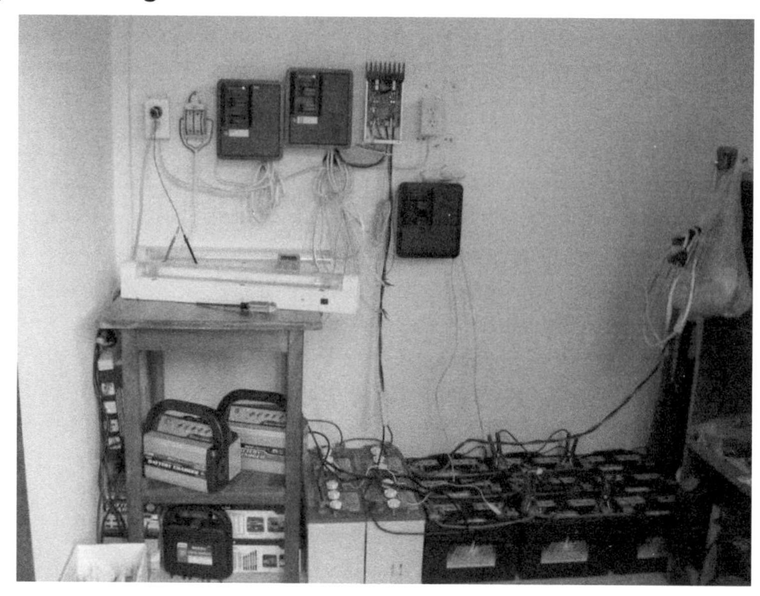

The energy room (above) has, (left to right) an outlet from the transfer switch to its right, shown in the up position selecting the generator outside. The load center for distribution is to its right. Next is the breaker box for the solar panels, then the solar charger (with front-panel removed). At the extreme right is another outlet from the generator conduit, going through the wall. Below it, and not wired into the system, is a breaker box to be inserted between the solar charger and the battery bank. It will allow the batteries to be switched out so that the charger battery terminals can be disconnected without the danger of shorting them.

A utility table is used for energy-room work. It contains a fluorescent light (2 shown) and a DMM for system monitoring and diagnosis. On its lower shelf are two off-line battery chargers, used when the generator is running. The inverter is behind them. To the right of the table, on the tile floor, is the battery bank - a fitting place for such heavy components!

Panning to the right of the above picture is the rest of the energy room, as shown below. With an entire room, shelving for storing spare electrical components (and other infrastructure parts) and tools are well-illuminated by the two windows in the room, one facing east, the other south.

Not all energy rooms will be this 'ideal'. This particular room was designed intentionally for electric system use as part of the house. Smaller spaces are quite adequate, though keep in mind that an electric system failure at an inconvenient time is best handled under favorable repair conditions.

Hopefully, this electric power facility will give you both ideas and encouragement in building your own home electric system.

Index

M

N

O

P

R

S

T

V

W